Chemometrics in Practical Applications

Chemometrics in Practical Applications

Editor
Adrianna Coty

Chemometrics in Practical Applications
Edited by **Adrianna Coty**

ISBN: 978-1-68117-198-2
Library of Congress Control Number: 2016934745

© 2017 by
SCITUS Academics LLC,
www.scitusacademics.com
Box No. 4766, 616 Corporate Way,
Suite 2, Valley Cottage,
NY 10989

This book contains information obtained from highly regarded resources. Copyright for individual articles remains with the authors as indicated. All chapters are distributed under the terms of the Creative Commons Attribution License, which permits unrestricted use, distribution, and reproduction in any medium, provided the original author and source are credited.

Notice

Reasonable efforts have been made to publish reliable data and views articulated in the chapters are those of the individual contributors, and not necessarily those of the editors or publishers. Editors or publishers are not responsible for the accuracy of the information in the published chapters or consequences of their use. The publisher believes no responsibility for any damage or grievance to the persons or property arising out of the use of any materials, instructions, methods or thoughts in the book. The editors and the publisher have attempted to trace the copyright holders of all material reproduced in this publication and apologize to copyright holders if permission has not been obtained. If any copyright holder has not been acknowledged, please write to us so we may rectify.

Preface

Chemometrics, the branch of science, is the use of mathematical and statistical methods to improve the understanding of chemical information and to correlate quality parameters or physical properties to analytical instrument data. It can be applied in predictive issues solving like predicting the target properties, desired features. It also can be used for the descriptive issue solving like the model composition, identification and understanding. Chemometrics shows its application in the multivariate data collection and analysis. Various algorithms and analogous ways are available for processing and evaluating the data. They can be implemented to various fields, like medicine, pharmacy, food control, and environmental monitoring. Chemometric techniques are particularly heavily used in analytical chemistry and metabolomics, and the development of improved chemometric methods of analysis also continues to advance the state of the art in analytical instrumentation and methodology. It is an application-driven discipline, and thus the standard chemometric methodologies are very widely used industrially, academic groups are dedicated to the continued development of chemometric theory, method and application development. Many chemical problems and applications of chemometrics involve calibration. The objective is to develop models which can be used to predict properties of interest based on measured properties of the chemical system, such as pressure, flow, temperature, infrared, Raman, NMR spectra and mass spectra. This book, Chemometrics in practical applications, presents several practical applications of chemometric methods in chemistry, biochemistry and chemical technology.

Table of Contents

Chapter 1	Chemometrics: Theory and Application	1
Chapter 2	Analysis of ATR-FTIR Absorption-Reflection Data from 13 Polymeric Fabric Materials Using Chemometrics	15
Chapter 3	Chemometric Analysis of the Amino Acid Requirements of Antioxidant Food Protein Hydrolysates	27
Chapter 4	QSAR Study of Antimicrobial 3-Hydroxypyridine-4-one and 3-Hydroxypyran-4-one Derivatives Using Different Chemometric Tools	43
Chapter 5	Comprehensive and Comparative Metabolomic Profiling of Wheat, Barley, Oat and Rye Using Gas Chromatography-Mass Spectrometry and Advanced Chemometrics	63
Chapter 6	Discrimination of Wild-Grown and Cultivated Ganoderma lucidum by Fourier Transform Infrared Spectroscopy and Chemometric Methods	81
Chapter 7	Spectroscopic Discrimination of Bone Samples from Various Species	99
Chapter 8	New Approachs in Drug Quality Control: Matrices and Chemometrics	113
Chapter 9	Application of Chemometrics to the Interpretation of Analytical Separations Data	123
Chapter 10	Chemometrics of Cells and Tissues Using IR Spectroscopy – Relevance in Biomedical Research	149
Chapter 11	Hyperspectral Imaging and Chemometric Modeling ofEchinacea — A Novel Approach in the Quality Control of Herbal Medicines	179

Chapter 12 Multielemental Composition of Suet Oil Based on
           Quantification by Ultrawave/ICP-MS Coupled with
           Chemometric Analysis   201

Chapter 13 Chemometric Feature Selection and Classification of
           Ganoderma lucidum Spores and Fruiting Body Using
           ATR-FTIR Spectroscopy   215

Chapter 14 Chemometric Analysis of an Sanitary Landfill
           Leachate   231

Chapter 15 Profiling of Fatty Acids Composition in Suet Oil Based
           on GC–EI-qMS and Chemometrics Analysis   243

           Index   259

CHAPTER 1

Chemometrics: Theory and Application

dos Santos Hilton Túlio Lima [1], Freitas Wagner [1,4], André Maurício de Oliveira[2] and de Melo Patrícia Gontijo [3]

[1] University of São Paulo (USP), Brazil
[2] Federal Center of Technology – Minas Gerais (CEFET - MG), Brazil
[3] Federal University of Uberlândia (UFU), Brazil
[4] State University of São Paulo (UNESP), Brazil

1. INTRODUCTION

This chapter aims to present a chemometrics as important area in chemistry to be able to help work with many among of data obtained in analysis. The term *chemometrics* was introduced in initial 70th years by SvantWold (Swede) and Bruce Kowalski (USA). According International Chemometrics Society, founded in 1974, the accept definition to chemometrics is (i) the chemical discipline that uses mathematical and statistical methods to design or select optimal measurement procedures and experiments (ii) to provide maximum chemical information by analyzing chemical data [1]. When the study involving many variable became the study in a multivariate analysis, so it is necessary to building a typical matrix and is normal to do a pre-processing. Pre-processing is a procedure to adjust the different factors with different units in values than allow give for each factor the same change to contribute to the model. After, next step is usually the Pattern Recognition method, to find any similarity in your data. In This method is common using the unsupervised group where there are the HCA and PCA analysis and the supervised group where there is the KNN. The HCA analysis (Hierarchical Cluster Analysis) is used to examine the distance among the samples in two dimensional plot (dendogram) and cluster samples with similarity. (Figure 1). Now PCA analysis (Principal Component analysis) is used to try decrease the size data set, without lost information about samples (Figure 2) and KNN used to classify samples using cluster previously know [2].

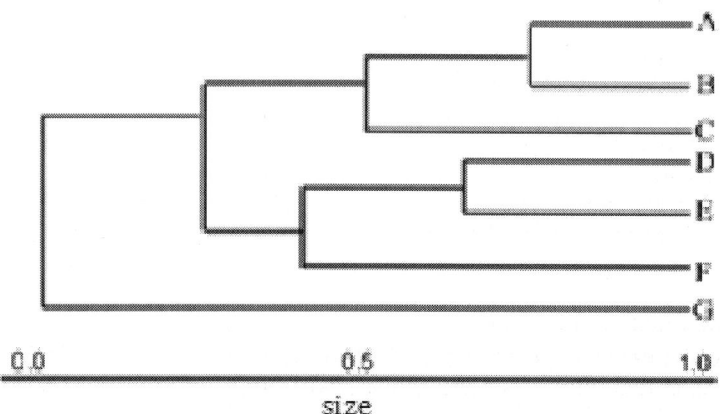

Figure 1. Example of dendogram

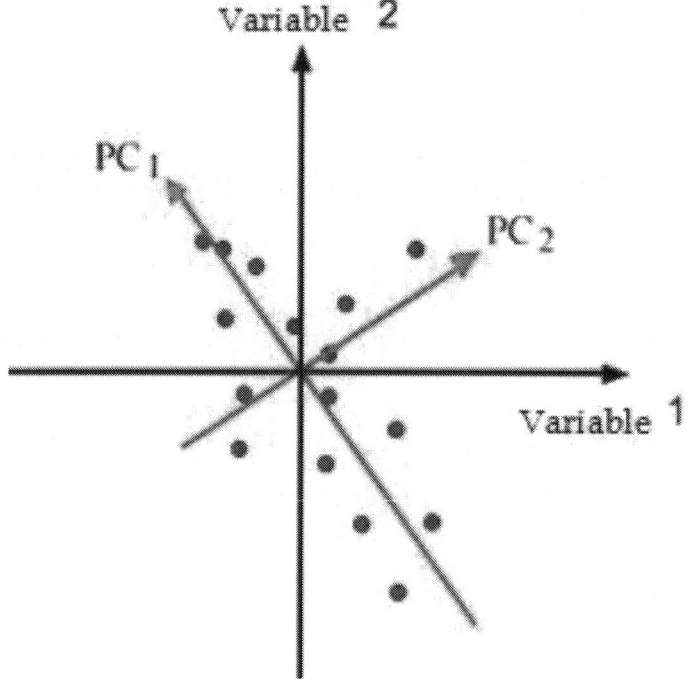

Figure 2. Clustering by PCA

Thus, the chemometrics show to be wide may be used in several area of knowledge.

2. PATTERN RECOGNITION

In analytical chemistry when we have the data set, it is important find similarities and differences between samples based on measurements. For this is necessary to use methods according with information about the samples. And can be: Unsupervised (HCA and PCA) and Supervised methods (KNN)

2.1. Unsupervised methods
In this group there are two methods: Hierarchical Cluster Analysis (*HCA*) and Principal Components Analysis (PCA), and the goal is to evaluate if there is any clustering in data set without using the class about samples.

2.1.1 Hierarchical Cluster Analysis *(Hca)*
The Hierarchical Cluster Analysis is a technique to evaluate the distance between de samples and group in a plot calling dendogram. Theses distance can be calculated utilizing different methods as Euclidean or Mahalanobis or Manhattan distance, for example. For the Euclidean distance is using theequation 1, for Mahalanobis distance is using the equation 2 and for Manhattan distance is usingequation 3:

$$\text{Distance} = \sqrt{(X_1 - Y_1)^2 + (X_2 + Y_2)^2 + \cdots + (X_n + Y_n)^2} \quad (1)$$

Where:

X_n and Y_n are the coordinates of sample X and Y in the nth dimension of row space.

$$\text{Distance} = \sqrt{(X_i - Y_j)^T C^{-1}(X_i + Y_j)} \quad (2)$$

Where:

X_i and Y_j are column vectors for objects *i* and *j*, respective and *C* is the covariance matrix.

$$\text{Distance} = \sum_{i=1}^{p} |X_i - Y_i| \tag{3}$$

Where:

X_i and Y_i are vectors.

When performed the estimate for distance, so is possible plot the dendogram. A general dendogram is showing below (Figure 3). In this dendogram is possible to see the samples (letters) and the distances (numbers). Samples belonging to clusters A, has a distance of 0,2 from one another. Same time the sample B has a distance 0,5 from cluster A. The value of distance can change according with the distance used to calculate.

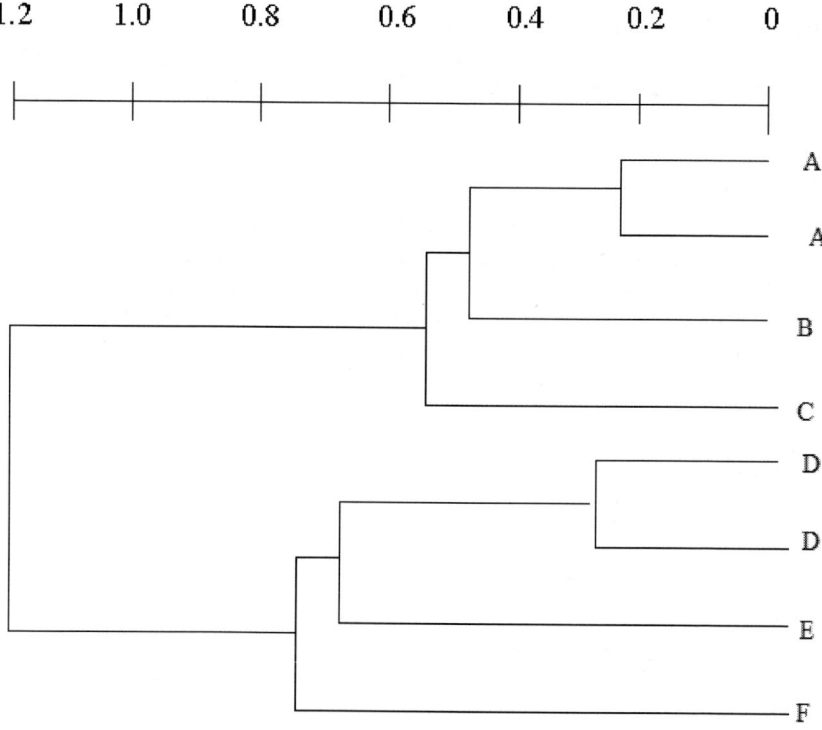

Figure 3. The general dendogram where above are the distances and right side are the samples

2.1.2. Principal Components Analysis (Pca)

The Principal Components Analysis (PCA) has the goal available the distances between the points using few axes in the row plot. In a matrix, each row is the point in the graphic below (Figure 2). So the aim is study the relationship between these samples to find the similarity and differences. In this general example are using two principal components (PC1 and PC2). The first PC (PC1) describes the major points in the graph and the maximum amount of variance, while the PC2 explain the remaining points. It is important to know that the sum of percentage described by PC´s must be close 100%. Another propriety of PC´s is about de position. The PC´s are always perpendiculars one with another.

The PCA technical can be used to define which variables are more important in a process. For this analysis is necessary use the factors (column in the matrix) and objects (row in the matrix). When the aim is to determine which variable are more important for the process is used *loading* and when want studying the relationship between objects is used *scores*

2.2. Supervised Methods

The Supervised methods are using when want to construct a model using the class membership for future samples. In this group, K-NN is a technical widely used when the goal is this.

2.2.1. K- Nearrest Neighbor (K-NN)

The KNN technical allows use the samples or clusters to identify another samples or clusters. For this is necessary to calculate the distances between them, using a Euclidean or Mahalanobis or Manhattan distance, for example. The minimum distance is calculated and the object is assigned to the corresponding class. A classification is dependent on the number of objects in each class.

3. CHEMOMETRICS IN MEDICINAL CHEMISTRY

3.1. The Qsar Principle: Hansch Analysis

The development of new drugs is a continuous challenge, before uncountable diseases the lack an adequate pharmaceutical approach. The modern medicinal chemists concern specially with methods based upon rational and quantitative procedures, aiming to focus on potentially efficient candidates. In that context, the use of chemometric methods is very important, in quantitative structure-activity relationship (QSAR) studies, and it presupposes that the biological activity (BA), measured through a biological response (BR), keeps a relationship with chemical structure (CS):

$$BR = f(CS) \tag{4}$$

The first attempt to quantitatively relate chemical structure to chemical behavior in a series of structuraly kindred compounds remounts to 1940's, with Hammett [3] who, studying the meta- and para-substituted benzoic acids at 25°C, stablished linear relationships between the R = X substituted benzoic acid ionization constant (KX) and the ionization constant of the non-substituted benzoic acid (R = H):

$$(m-/p-R)C_6H_4COOH \rightarrow (m-/p-R)C_6H_4COO^- + H^+ \tag{5}$$

$$\sigma = \log\left(\frac{K_X}{K_H}\right) = \log(K_X) - \log(K_H)$$

The σ constant is group-specific, and represents the electronic effect (inductive and resonance type) pursuit by R group. In 1964, Corwin Hansch [4] combined the use of the electronic constants to the lipophylic parameter (π), which represents the contribution of each R group to the overall lipophylicity:

$$\pi = \log\left(\frac{P_X}{P_H}\right) = \log(P_X) - \log(P_H) \tag{6}$$

where P_X is the X-substituted compound octanol-water partition coefficient, and P_H, the partition coefficient for a non-substituted compound. Thus, a QSAR equation evolves some kind of RB, for example, the negative logarithm of the minimal inhibitory concentration (MIC) for am antimicrobial compounds series (-log(MIC)), and the electronic (σ) and lipophyliceffect (π) of the R groups, the makes distinction among the several series representatives, can be expressed as

$$-\log(MIC) = \log\left(\frac{1}{MIC}\right) = a \cdot \sigma + b \cdot \pi + c \tag{7}$$

where a, b and c are the multiple regression coefficients.

The Hansch's hypothesis that RB may be related to specific physico-chemical to each substituent present in the basic skeleton in a congener series of similar BA led to the proposition of numerous descriptors, of different kinds, useful to the identification of the principal effects that show up in drug action.

3.2. Physico-Chemical Descriptors

There are several physico-chemical descriptors, useful in QSAR studies that can be divided in categories: constitutional, topological, stereochemical and electronic ones, beside the so called indicator variables.

3.2.1. Constitutional descriptors
This kind of descriptor is related to the presence of structural characteristics that can affect the BA, such as: amount of unsaturated bonds, amount of hydrogen-bond donors, average ring size, etc.

3.2.2. Topological descriptors
These are descriptors that represent shape and connectivity, such as: ramifications, spacing groups, unsaturations, etc. The Kier [5] and Wiener [6] descriptors are typical.

3.2.2. Steric (or stereochemical) descriptors
Steric descriptors exist to describe effects related to the size of chemical groups and hindrance behavior. Taft steric descriptor, Es,[7] is a common example.

3.2.4. Eletronic descriptors
These variables are related to molecular electronic densities, and are used to be calculated by quantum methods. One can mention as examples: dipole moments, atomic partial charges, highest occupied molecular orbital energy (HOMO) and lowest unoccupied molecular orbital energy (LUMO).

3.2.5. Indicator variable and Taylor analysis
Indicator variables represent a useful way to convert a qualitative information into quantitative once, just as the occurrence of some kind of structural feature – setting 1 when this feature is present, and 0 otherwise. The Taylor QSAR [8] approach employs indicator variables.

3.3. Chemometric Methods Applied To Drug Design

Chemometric statistical methods find in QSAR a large application field, considering that the multivariate problems are inherent to it.

3.3.1. Discriminatory and classificatory methods
Those methods aim the grouping and classification of compounds and variables in classes or categories that share resemblances, and are very interesting in pattern recognition situations and in dimensionality reduction of complex systems.

3.3.2. Principal Component Analysis (PCA)
Principal component (PCs) methods aim to combine correlated variables, projecting them in a new coordinate system, so that fewer variables are obtains, without any intercorrelation. The former coordinates are projects in a new axis

system, in which the system variability is maximum along PC1, decreasing along the other axises (PC2, PC3...), all of the orthogonal each other, what allows one to deal just with the first components (usually PC1, PC2 and PC3). Thus, from a multi-variable universe, commonly multicolinear, one can obtain a simpler system with almost the same amount of information. Naming X the data matrix, with I×J dimension (I molecules and J descritors), a PCA generates two matrices, T e L, so that

$$X = TL^T \tag{8}$$

The matrix T is of scores, and represents the position of the compounds in aa novel coordinate system in which the components are its axises, and L is the loading matrix. Plotting the PCs instead of the original descriptors, one obtains groups governed by the similarities among the data.

3.3.3. Hierarchical Cluster Analysis (HCA)
This analysis is also useful to the classification of compounds, permitting visually distinguish the patterns and cluster. The plot resembling a tree, called dendogram, presents similar compounds at the same branches. Those branches are plotted based upon a similarity matrix, S, and each component of it is given by the similarity index between two samples k and l, S_{kl}:

$$S_{kl} = 1.0 - \frac{d_{kl}}{d_{max}} \tag{9}$$

In this expression, d_{kl} is the Euclidian distance between k and l, and d_{max}, the maximum distance. Ferreira [9] describes a PCA/HCA analysis for a 25-compound series of 1,4- naphtoquinones with antitumour activity. Using electronic descriptors, it was possible to distinguish active from inactive compounds (Figure 4). The loadings values indicate that the presence of high-density groups in side chain and terminal positions favours activity. The same profile arise from the dendogram analysis.

3. CHEMOMETRICS IN MEDICINAL CHEMISTRY

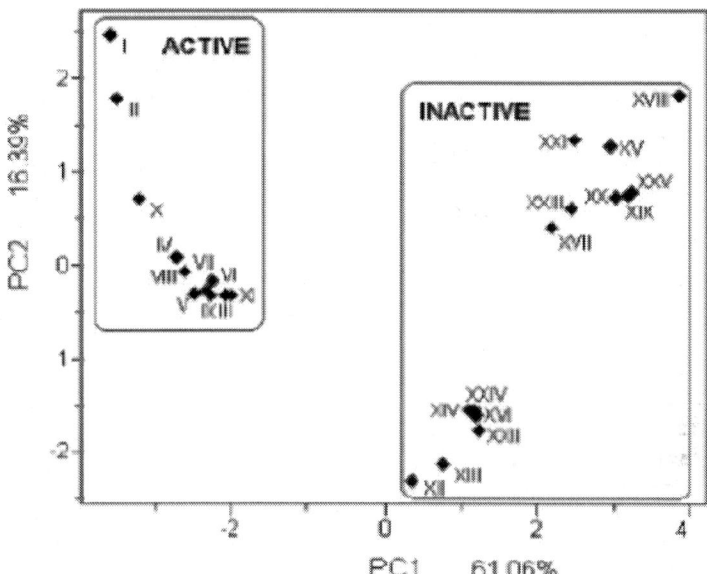

Figure 4. PC1 versus PC2 scores plot.

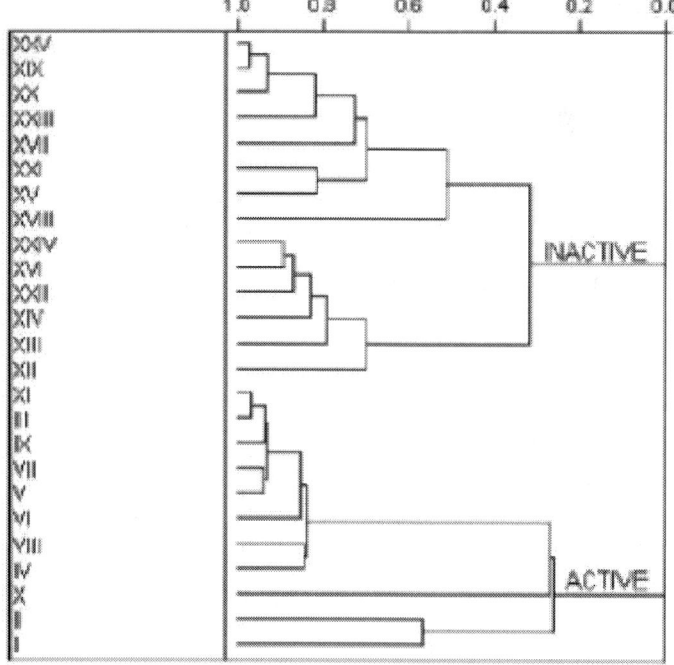

Figure 5. Dendogram for a naphtoquinone series

3.4. Multivariate Regression

To construct a QSAR equation (Eq. 1), it is necessary to adopt some kind of multivariate fitting method in order to correlate the descriptors with the BR. The main methods are: multilinear regression (MLR), principal component regression (PCR) and partial-least squares (PLS).

3.4.1. Multilinear regression (MLR)

The objective of this method is obtaining a relationship among a number of descriptors limited to 1/5 of the number of compounds and the BR, as an equation of the form:

$$BR = \alpha_1(\pm\varepsilon_1) \cdot D_1 + \alpha_2(\pm\varepsilon_1) \cdot D_2 + \alpha_3(\pm\varepsilon_1) \cdot D_3 + \cdots + \varepsilon \quad (10)$$

in which i are the regression coefficients, Di are the descriptors, εi, the coefficients confidence interval and ε, the independent term. The model statistical validation is very important, and it requires the consistency in the Di descriptors unit, as well as in values magnitude (necessarily,). Statistical parameter like the fitting coefficient (r), the sample standard deviation (s), the cross-validation coefficient (q^2) and the Fischer test (F) are used in this task. The MLR is quite sensitive to multicollinearity: variables intercorrelated (tipically, com $r^2 > 0.6$) must not be used together. This is a common problem in multi-descriptor system that may be dealed with other regression methods.

3.4.2. Principal component regression (PCR)

In order to avoid multicollinearity, it is possible to make the regression, not with the descriptors themselves, but with their principal components (PCs) generated in a PCA treatment. The main advantage of this approach is the assurance that every variable are independent and no n-correlated, despite it is necessary to analyze the loading matrix (L). In this kind of regression, the variables are defined to maximize the descriptor matrix variance, without force a correlation with the BR

3.4.3. Partial Least Square (PLS)

Similarly to PCR, the PCs are employed, but in this case, the BR matrix has maximum variability, so that each loading matrix component (L) is a good predictor for each BR matrix component. This is the most used regression method, and it is adequate for dealing with 3D-QSAR problems, in which a set of compounds preciously aligned is put within a grid of interaction points with a molecular probe. Each point energy is a variable in the QSAR equation, which are by their turn corrlated with the BR to achieve a tridimensional profile of the critical sites that favours or disfavours the interaction with a hypothetical biological receptor.

4. DESIGN OF EXPERIMENTS

The exploration for new sources of energy such as biodiesel is of great importance today as well as their production processes. The factorial design is an important tool to reduce the search time, waste of reagents and hence operating costs [10]. A factorial design is performed with the interest to determine the experimental variables and interactions between variables that have significant influence on the different responses of interest [11]. After selecting the significant variables, we must evaluate the experimental methodology and the influence of a particular variable on the yield of the reaction, a statistical experimental design, full factorial type, in which the independent variables are: the nature and concentration of catalyst temperature and the molar ratio between alcohol and oil and the dependent variable is the yield of esters produced. The variables that were not selected must be fixed throughout the experiment [12]. In a subsequent step must be chosen which planning used for estimating the effect (the effect) of the different variables results in a reduced number of conducting experiments. In the screening study the interactions between the variables (main interactions) and second order, usually obtained by full or fractional factorial designs. In the experiments are evaluated best experimental conditions, as well as their simultaneous effects that influence the yield of the reaction are therefore extremely important for understanding the behavior of the system [13]. The values of "p" and greater than or equal to 0.05 indicate that the factors: variable (1), variable (2), variable (3), variable (4) and the interactions of the variables are statistically significant at 95% reliable, since they are greater than 0.05. These parameters were evaluated at a low level (-1) and high (+1) are significant to the process of positive or negative manner. The Figure. 6 shows the profile of the Pareto chart [7].

The analysis parameters obtained by means of multivariate optimization consists in choosing the conditions for preliminary assessment of experimental variables (fractional factorial design) followed by a response surface methodology (central composite design) made from the screening of the variables that may affect the synthesis of biodiesel. Generated model and the set of significant effects can evaluate through the study of response surface methodology, as shown in Figure 7 and 8, and their interference in the response, ie the yield of the reaction, in which the dark area demonstrates the conditions that process has higher yield.

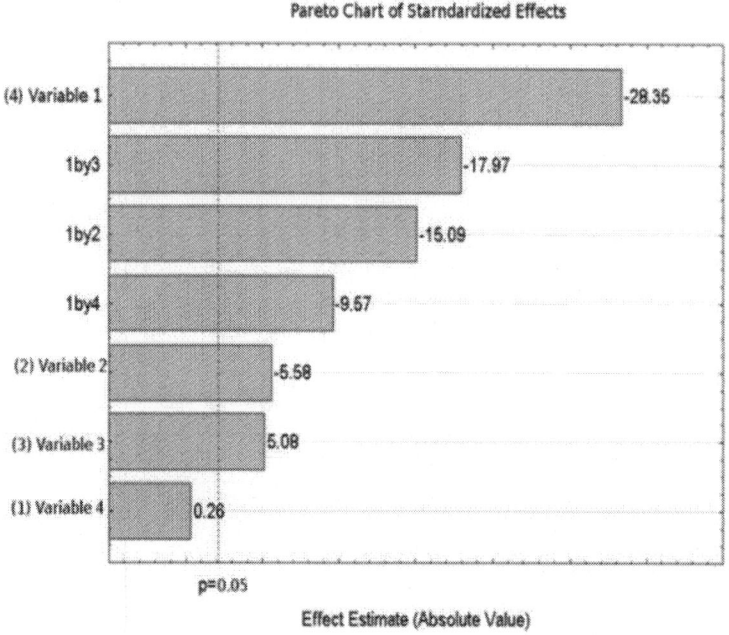

Figure 6. Pareto chart of the resulting fractional factorial design to evaluate the effects of each variable and their interactions in the reaction yield.

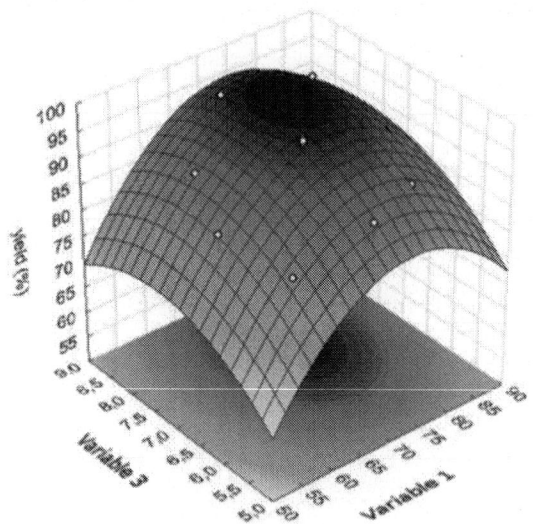

Figure 7.A) Response surface generated by the central composite design for optimization of variables 1 and 3

Thus, the statistical analysis shown to be an important tool to evaluate, select and propose new technological routes, either through raw materials and / or process evaluation of the parameters that most influence the transesterification reaction to obtain for biofuels.

5. CONCLUSION OF CHAPTER

This chapter had as aim to show the versatility tools chemometrics in several areas. Was showed application chemometrics theory in drug design, natural products chemistry but it is not limited in theses area. Well, we hope to have expanded the range of chemometrics

REFERENCES

1. M Otto, Chemometrics- Statistic and Computer Application in Analytical Chemistry. Ed. Wiley-VCH. 1999
2. K Beebe, R Pell, M Seasholtz, Chemometrics- A Practical Guide. Ed.WileyInterscience Publication. 1998
3. HammettLouis 1937J. Am. Chem. Soc. 59: 96.
4. C Hansch, 1969A Quantitative Approach to Biochemical Structure-Activity RelationshipsAcct. Chem. Res. 2232239
5. HallLowell H.; Kier, Lemont B. (1976Molecular connectivity in chemistry and drug researchBoston: Academic Press.
6. H Wiener, 1947Structural determination of paraffin boiling points"J. Am. Chem. Soc. 1691720
7. R. W Taft, Linear free energy relationships from rates of esterification and hydrolysis of aliphatic and ortho-substituted benzoate esters. J. Am. Chem. Soc. 195219527427292732
8. C Hansch, P. G Sammes, J. B Taylor, Comprehensive medicinal chemistry: the rational design, mechanistic study & therapeutic application of chemical compounds,Pergamon Press: Oxford, 19904
9. M. M. C. J Ferreira, Braz. Chem. Soc., 1362002
10. M Charoenchaitrakool, J Thienmethangkoon, 2011Statistical optimization for biodiesel production from waste frying oil through two-step catalyzed processFuel Processing Technology112 EOF118 EOF
11. M Berrios, M. C Gutiérrez, M. A Martín, A Martín, 2009Application of the factorial design of experiments to biodiesel production from lardFuel Processing Technology1447 EOF1451 EOF

12. P. G Melo, 2012Production and characterization of obtained from Macaúba (Acrocomiaaculeata). Master degree thesis. Univesity of Federal of Uberlandia- Brazil.
13. I. M Atadashi, M. K Aroua, A. A Aziz, 2010High quality biodiesel and its diesel engine application: A reviewRenewable and Sustainable Energy Reviews1999 EOF2008 EOF
14. S. A Mingoti, 2007Data analysis through methods of multivariete statistical approach applied.Federal University of Minas Gerais

CHAPTER 2

Analysis of ATR-FTIR Absorption-Reflection Data from 13 Polymeric Fabric Materials Using Chemometrics

Innocent Pumure[1]*, Shannon Ford[1], Jessica Shannon[1], Christopher Kohen[1], Amanda Mulcahy[1], Kelvin Frank[1], Sheri Sisco[1], Nhamo Chaukura[2]

[1]College of Heath, Science and Technology, School of Environmental, Physical and Applied Sciences, University of Central Missouri, Warrensburg, MO, USA
[2]Department of Polymer Science & Engineering, Harare Institute of Technology, Belvedere, Harare, Zimbabwe

ABSTRACT

We used both correlation and covariance-principal component analysis (PCA) to classify the same absorption-reflectance data collected from 13 different polymeric fabric materials that was obtained using Attenuated Total Reflectance-Fourier Transform Infrared spectroscopy (ATR-FTIR). The application of the two techniques, though similar, yielded results that represent different chemical properties of the polymeric substances. On one hand, correlation-PCA enabled the classification of the fabric materials according to the organic functional groups of their repeating monomer units. On the other hand, covariance-PCA was used to classify the fabric materials primarily according to their origins; natural (animal or plant) or synthetic. Hence besides major chemical functional groups of the repeat units, it appears covariance-PCA is also sensitive to other characteristic chemical (inorganic and/or organic) or biochemical material inclusions that are found in different samples. We therefore recommend the application of both covariance-PCA and correlation-PCA on datasets, whenever applicable, to enable a broader classification of spectroscopic information through data mining and exploration.

Keywords: Data Mining, Principal Component Analysis, ATR-FTIR, Polymeric Materials, Chemometrics

1. INTRODUCTION

An isolated non-linear molecule containing N atoms can undergo 3N-6 normal vibrational modes each often consisting of complex mixture of bond stretches and other deformations. In a polymeric substance, vibrational modes of a polymer depend on the chemical nature of the repeating monomer units. The resulting frequencies and intensities of the vibrational modes are sensitive to chemical structure and therefore provide essential information that can be used to characterize chemical substances [1]. ATR-FTIR is a technique that can be used to collect absorption-reflection data that is resonant with allowed vibrational transitions of a variety of chemical functional groups. Analysis of the collected data can be done using univariate or selective group frequency assignments or by means of multivariate statistical methods of analysis such as Principal component analysis (PCA). Martin et al. [2] used FTIR and covariance-PCA to classify vegetable oils. Kraft et al. [3] used first derivatives of normalized spectra to classify different polymers using spectral imaging in an online process. Brereton et al. [4] applied dynamic mechanical analysis of polymer properties as temperature is changed. The method involved chemometric analysis of the damping factor (tan δ) as a function of temperature. It was demonstrated that thermal analysis together with chemometrics provides excellent discrimination, representing an approach for characterization of polymers [4] -[6].

PCA is an unsupervised non-parametric multivariate statistical technique that is frequently used in data mining to unveil or confirm relationships and similarities between samples and measured parameters. PCA is used to reduce data dimensionality [7]-[11] by identifying similar patterns between samples and introducing new latent variables which are linear combinations of the measured variables. This allows for pattern recognition and streamlining of large data sets to smaller datasets while still retaining the original chemical meaning of the data. Correlation PCA begins with normalization followed by standardization of data to give appropriate weighting. On the other hand covariance PCA involves standardization of data matrix excluding the normalization step. Standardization is the transformation of measured data to give a mean of zero and a standard deviation of 1. Hence standardization transforms all data into z scores. Prior to normalization and standardization the raw data can be pre-processed by performing preliminary transformations that include taking logarithms, inverses, squares or square roots of the raw data. Z scores will then be obtained from the standardization of the transformed but related data. A plot of the z scores between any two data sets will lead to existing or new linear relationships/combinations or clustering of data points of samples with similar attributes. The new linear combinations are known as principal components (PCs). The generated PCs are usually orthonormal to each other. Placing a meaning on the derived PCs or clusters requires a thorough knowledge of the chemistry and the physical properties of the samples analyzed. There may be prior or apriori knowledge about the clustering or data classification before the application of PCA.

Ideally, combined variance from all PCs used should explain at least 80% of the total variance observed. However in most situations the first two PCs that explain the largest total variance are frequently used to interpret data. A plot of the PCs over the sample space is called a Score plot which shows physicochemical connectivities between the samples. A plot of PCs over the variable space is called a Loading plot that yields connectivities among the variables. Superimposition of a score plot onto a Loading plot is known as a biplot which often provide direct details on the relationships between samples and variables. In this study, we show the difference in the physicochemical information that is obtainable by applying both correlation and covariance PCA to the same set of data obtained using ATR-FTIR spectroscopy for 13 different polymeric fabric materials.

2. MATERIALS

2.1. Fabric Materials

Fabric materials were obtained from Test Fabrics Inc. and were analyzed without any pretreatment. No sample pretreatment was considered necessary to be done before analysis. A swatch of the polymeric materials is shown in Figure 1. It appears, the polymeric binder material that runs perpendicular to the fibers of interest did not affect classification using PCA. Inspection of the ATR-FTIR spectra showed the expected peaks that are attributable to the major functional groups in the repeating monomer units of the polymers. This shows the robustness of the analytical method and PCA methods used.

2.2. Nicolet ATR-FTIR Instrument

The instrument was set to record absorption-reflection data for the fabric samples from 525 to 4000 cm^{-1} spectral region. Instrument parameters include the following; number of averaged scans 32, resolution 2, data spacing 0.964 cm^{-1} and Happ-Genzel signal apodization. A background scan was always recorded before performing any analysis. When a sample is analyzed, the Nicolet ATR-FTIR instrument used records a ratio between analytical signals and background signals at each wavenumber to counter variations in source intensity. A thin film of polystyrene was used for quality control (QC) and quality assurance (QA) purposes. QC data was collected and analyzed before samples were analyzed. An average score of 98.17% ± 1.32% of the polystyrene QC standard obtained against an internal reference Styrofoam standard was considered to be satisfactory. The 1.32 value corresponds to 3 standard deviations of the average score. The instrument was considered ready for use only after quality control conditions were met.

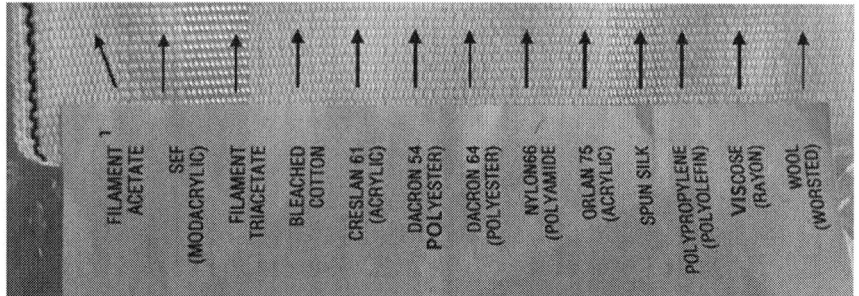

Figure 1. A strip containing all 13 different fabric materials as obtained from Test Fabrics Inc. The chemical formula of the repeating units of the polymers are shown in Table 1.

2.3. Analytical Procedure

Each set of fibers from a single type of a fabric material is held together by fibrils of a binder material that weave in a perpendicular direction. After the quality assurance step, ATR-FTIR spectra were recorded on each sample of the fabric material in the presence of the binder. The spectra for each individual fabric materials were run for a total five times, each spectra being recorded on a different spot. Absorption-reflection data from five runs on each fabric material were averaged and plotted in Figure 2. The ATR-FTIR data were then exported into Minitab 15.0 software for PCA analysis. Both covariance and correlation PCA techniques were then applied to the absorption-reflection data. Results from PCA are plotted in Figure 3(a) (correlation PCA), Figure 3(b) (single linkage-correlation coefficient dendrogram) and Figure 4 (covariance PCA) respectively.

3. RESULTS

In Figure 2, the data appears clouded, unclear and the underlying relationships between the 13 different polymeric fabrics cannot be easily identified just by analyzing the ATR-FTIR data simultaneously. Even though one can use selective organic functional group frequencies to identify the nature of the individual fabric materials, it appears challenging to succinctly classify the different fabric materials. However, the data can be collectively analyzed using PCA, a multivariate statistical technique.

Interpretation of PCA data requires a thorough understanding of the chemical properties of the analyzed samples. For a regular polymer chain, vibrations occurring in the repeating monomer units result in infrared absorption-reflection bands that can be used to characterize the polymer materials. The repeating monomer units and their corresponding functional groups in the 13 polymer materials are presented in Table 1.

3.1. Correlation-PCA

Analysis of the data matrix using correlation PCA resulted in the classification of polymers materials according to their major chemical functional groups. Similar classification was also found using the single linkage-corre- lation coefficient dendrogram. This information is shown in Figure 3(a) andFigure 3(b). Classification using correlation PCA is so powerful that polymers having identical or similar major chemical functional groups in their repeating monomer units are clustered together even though they might have different chemical origins. Although correlation does not necessarily mean causation [5], the clustering of the fabric materials on the PCA plot can be related to chemical properties of the fabric materials. All characteristic group frequencies for organic functional groups were obtained from Principles of Instrumental Analysis, 7th edition by Skoog, Holler and Crouch [12].

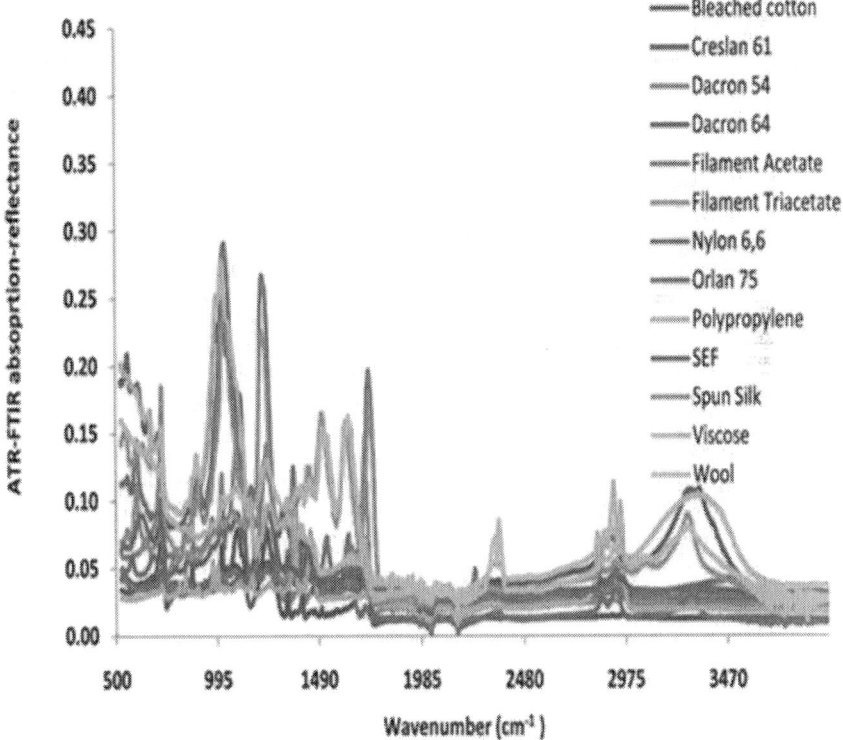

Figure 2. ATR-FTIR data for 13 different fabric materials.

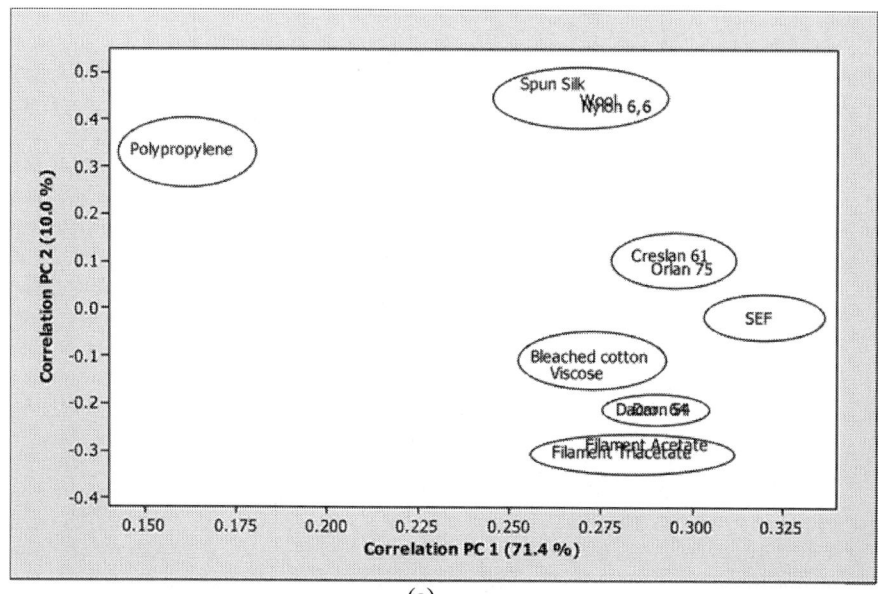

(a)

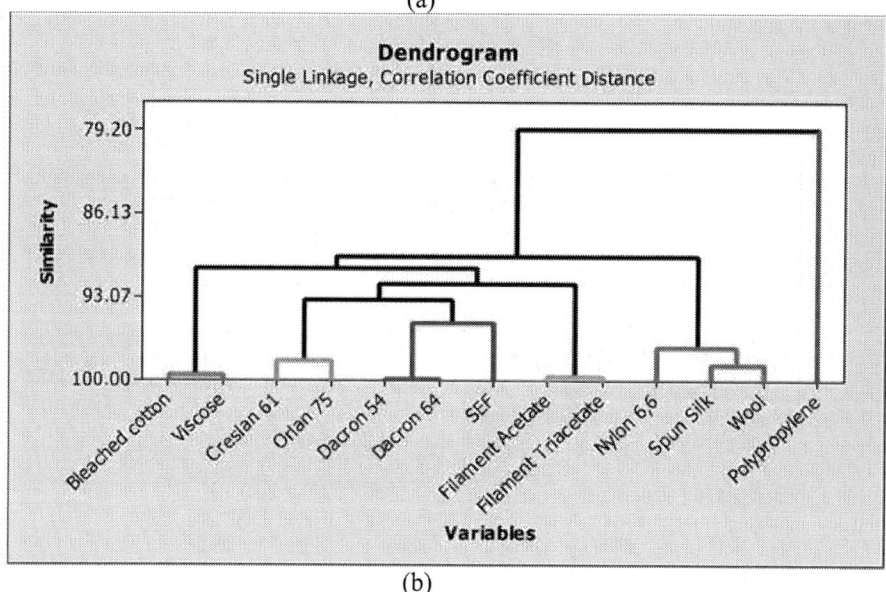

(b)

Figure 3. (a) Data from correlation PCA; (b) single linkage-correlation coefficient dendrogram.

3. RESULTS

Table 1. Repeating monomer units in the polymers.

Polymer	Type	Repeating monomer units
Cotton and Viscose	Cellulosic	
Filament acetate and Filament Triacetate	Acetylated cellulose $R = CH_3COO^-$	
Dacron 64 and Dacron 64	Polyester $R = [-CH_2-]_n$	
Silk, Wool and Nylon 6.6	Polyamides R_1 and R_2 are organic functional groups	
Creslan 61, Creslan 64 and SEF	Polyacrylonitrile	
Polypropylene	Polyolefinic	

Spun silk, Wool and Nylon 6,6 are condensation polyamide polymers. They all have the polyamide backbone in spite of having different three dimensional structural properties. The amide group is characterized by absorption reflection intensities for C-N at 1223, 1229 and 1243 cm^{-1} and for N-H at 3268, 3251 and 3289 cm^{-1} and C=O at 1620, 1633 and 1629 cm^{-1} for Spun silk, Wool and Nylon respectively. Spun Silk and wool are derived from animal protein, a natural origin, but Nylon 6,6 is a synthetic polyamide. Correlation PCA grouped these three polymers together most likely because they share a common amide functional group in their repeating monomer units.

Dacron is essentially a poly (ethylene terephthalate) polyester [13]. The two varieties of Dacron polymers, Dacron 54 and Dacron 64, are both synthetic polyesters. The two share an ester functional group and that could be the primary reason why they were grouped together using correlation PCA. The characteristic vibrational stretch frequencies associated with ester functional group are 1083 cm^{-1} (Dacron 54) and 1085 cm^{-1} (Dacron 64) for the C-O-C ether linkage and with 1713 cm^{-1} (Dacron 54) and 1709 cm^{-1} (Dacron 64) for the carbonyl group.

Orlan 75 and Creslan 61 represent different versions of polyacrylonitrile copolymers. They are made from the polymerization of acrylonitriles. The characteristic vibrational modes of the nitrile functional groups in the two polymers were found at 2237 cm^{-1} in Orlan 75 and 2241 cm^{-1} in Creslan 61. Poly acrylics usually contain at least 85% of polyacrylonitrile as acrylic fibers.

The two were clustered together using correlation PCA most likely because they are primarily made from polyacrylonitrile monomer.

SEF (Self Extinguishing Fiber) is a polyacrylonitrile copolymer that is mixed with other polymers. SEF did not group well together with other acrylics (Orlan 75 and Creslan 61) probably because of differences in composition and the chemical nature of additional copolymers used. SEF belongs to a class of polymers that are called modacrylics. Modacrylics typically contain 35% - 85% polyacrylonitriles as acrylic fibers and the remainder is added to deliberately fine tune the overall polymer properties. However, the positioning of SEF on the sample score plot (Figure 3(a)) and on the correlation dendrogram (Figure 3(b)), is not very far away from the two other acrylics and from Dacron 54 and Dacron 64. However, Figure 3(b) shows that there is a strong similarity in composition between SEF and the polyesters. In addition to a nitrile peak at 2239 cm^{-1} in SEF ATR-FTIR spectrum, there is a presence of relatively strong peaks at 1082 cm^{-1} and 1711 cm^{-1} that correspond to ether linkage and carbonyl group vibrational frequencies which provides evidence for the presence of a polyester.

Viscose and bleached cotton are both cellulosic because they are derived from plant material, a natural origin. Bleached cotton is made from cotton that has been reacted with chemicals which include oxidizing agents such as hydrogen peroxide or hypochlorite to remove color. Rayon is produced by the chemical treatment of cellulosic material using NaOH and CS_2 [14]. Both Bleached cotton and viscose essentially exhibit the same chemical properties. Similarity in chemical composition could be the reason why the two were grouped together using correlation PCA. Vibrational frequencies for hydroxyl groups were found at 3299 cm^{-1} in viscose and 3258 cm^{-1} in wool.

Both filament acetate and filament triacetate are made from the esterification of hydroxyl groups on cellulose using reagents such as acetic anhydride [15]. Acetylation of one hydroxyl group results in the formation of filament acetate and acetylation of all available three hydroxyl groups yields filament triacetate. Although they are derived from cellulose, they now have additional acetyl groups that are added onto the cellulosic polymeric chain. The two polymers are examples of modified natural polymers. The chemical composition of both Filament acetate and Filament Triacetate is now different from cotton and viscose due to acetylation. The characteristic peaks at 1727 cm^{-1} (Filament acetate) and 1731 cm^{-1} (Filament triacetate) indicate a strong presence of the carbonyl functional groups added through acetylation. There is also an apparent characteristic decrease in the relative intensities of OH (around 3300 cm^{-1}) functional group due to acetylation.

Polypropylene is an addition polyolefinic polymer that is made up of a repeating unit containing saturated hydrocarbons. Polypropylene is the only polyolefinic polymer used in the study and that is why it is well separated from the other polymers as shown in Figure 3(a) and Figure 3(b). Polypropylene is also a synthetic polymer. The C-H in the repeating units have characteristic ATR-FTIR bands at 2915, 1376 and 1450 cm^{-1}.

3.2. Covariance PCA

Covariance PCA on the other hand was used to provide additional information regarding other properties of the fabric materials not explained using correlation PCA. Covariance PCA is normally applicable to data measurements performed using a similar scale, more specifically if only one instrument is used. Covariance PCA enabled the classification of polymeric fabric materials primarily by origin and partially by functional groups. This information is shown in Figure 4. Natural impurities, residual synthetic chemicals, and additives [16] present in the fabric materials could have played a critical role in enabling the covariance PCA to classify the fabric materials according to their chemical origins. It appears natural inclusions from animal origins enabled the spun silk and wool to be grouped together. Both spun silk and wool are polyamides that have a natural animal origin. However, even though Nylon 6,6 is a polyamide that is chemically similar to wool and silk (Figure 3(a) and Figure 3(b)), it was grouped together with the other purely synthetic polymers by covariance PCA. It was really interesting to note that Nylon 6,6, a synthetic polymer, grouped with other synthetic polymers and not with other polyamides of animal origin after the application of covariance PCA.

Bleached cotton and Viscose both originate from cellulose. Results from covariance PCA and Correlation PCA are similar for the two primarily because their origin and chemical properties are essentially the same [17]. Filament acetate and Filament triacetate originate from plant materials but they have additional acetylated groups that make them different from bleached cotton and viscose.

Nylon 6,6, polypropylene, SEF, Orlan 75, Creslan 61, Dacron 64 and Dacron 54 are all synthetic polymers. Their extended group (Figure 4) is conspicuously far away from those with natural origins. Hence correlation PCA enabled the synthetic polymers to be grouped most likely because of the synthetic residual chemical inclusions and other additives [18]. The closeness of SEF to Dacron 54 and Dacron 64 in both correlation and covariance PCA data provides information that the polyacrylamide in SEF could have been blended with a polyester material.

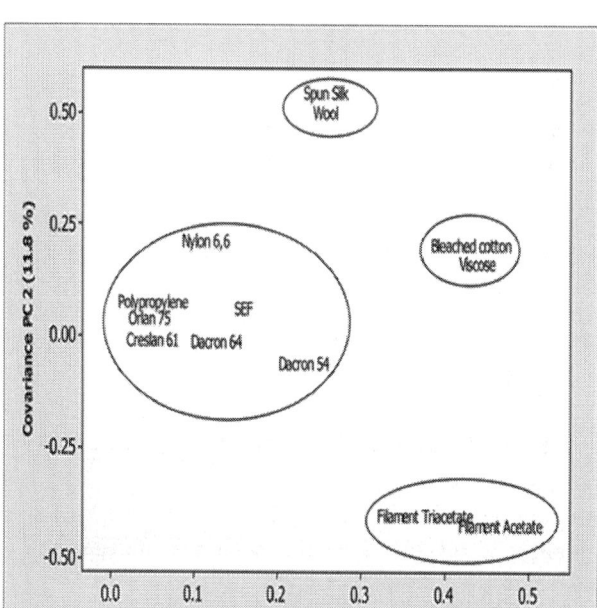

Figure 4. Data from covariance PC.

4. CONCLUSION

Application of both correlation PCA and covariance PCA to chemical data allows for a better interpretation of chemical and biological data hidden in a maze of ATR-FTIR data. Correlation PCA offered a way of classifying the polymers according to their chemical properties whereas covariance PCA was better at classifying the polymers according to their chemical or biological origins. Covariance PCA was able to separate Nylon 6,6 from Spun Silk and Wool because Nylon 6,6 is a synthetic polymer whereas Spun Silk and Wool are derived from animals. However, correlation-PCA was able to group all three together because they share a common polyamide backbone. The robustness of both PCA methods is also highlighted because the analyses of all the fabric materials were performed in the presence of a binder, but the binder material did not affect the chemometric classification process. Cserhati and Zoltan [19] suggested that the application of correlation matrix as a basis for the PCA may lead to slightly distorted results that strongly advocates for the use of covariance matrix in PCA. In this study we showed that the two techniques, correlation-PCA and covariance PCA complement each other and therefore are equally important. It is therefore highly encouraged that both PCA methods be applied in data analysis whenever they are both applicable to allow for a holistic approach towards data classification and interpretation.

ACKNOWLEDGEMENTS

The authors wish to thank the University of Central Missouri for providing financial support and the Test Fabrics Inc. for providing the fabric material samples.

REFERENCES

1. Hunt, B.J. and James, M.I. (1993) Polymer Characterization. Blackie Academic and Professionals, New York and London.
2. Rusak, D.A., Brown, L.M. and Martin, S.D. (2003) Classification of Vegetable Oils by Principal Component Analysis of FTIR Spectra. Journal of Chemical Education, 80, 541.
3. Kulcke, A., Gurschler, C., Spöck, G., Leitner, R. and Kraft, M. (2003) On-Line Classification of Synthetic Polymers Using Near Infrared Spectral Imaging. Journal of Near Infrared Spectroscopy, 11, 71-81.
4. Lukasiak, M.B., Faria, R., Zomer, S., Brereton, R.G. and Duncan, J.C. (2006) Pattern Recognition for the Analysis of Polymeric Materials. Analyst, 131, 73-80.
5. Lloyd, G.R., Brereton R.G., Faria, R. and Duncan, J.C. (2007) Learning Vector Quantization for Multiclass Classification: Application to Characterization of Plastics. Journal of Chemical Information and Modeling, 47, 1553-1563.
6. Fari, R., Duncan, J.C. and Brereton R.G. (2007) Dynamic Mechanical Analysis and Chemometrics for Polymer Identification. Polymer Testing, 26, 402-412.
7. Lavine, B.K. and Workman, J. (2013) Chemometrics. Analytical Chemistry, 85, 705.
8. Eriksson, L., Trygg, J. and Wold, S. (2014) A Chemometrics Toolbox Based on Projectionsand Latent Variables. Journal of Chemometrics, 28, 332-346.
9. Pienta, N.J. (2013) Taking a Cue from Statistics Education. Journal of Chemical Education, 90, 525.
10. Sandasi, M., Kamatou, G.P.P., Gavaghan, C., Barasnka, M. and Viljoen, A.M. (2011) A Quality Control Method for Geranium Oil Based on Vibrational Spectroscopy and chemometric Data Analysis. Vibrational Spectroscopy, 57, 242- 247.
11. Pumure, I., Renton, J.J. and Smart, R.B. (2011) The Interstitial Location of Selenium and Arsenic in Rocks Associated with Coal Mining Using Ultrasound Extractions and Principal Component Analysis (PCA). Journal of Hazardous Materials, 198, 151-158.

12. Skoog, D.A., Holler, F.J. and Crouch, R.S. (2007) Principles of Instrumental Analysis. 6th Edition, Chapter 17, Thompson Brookes/Cole, California, 461-463.
13. Kannan, R.Y., Salacinski, H.J., De Groot, J., Clatworthy, I., Bozec, L., Horton, M., Butler, P.E. and Seifalian, A.M. (2005) The Antithrombogenic Potential of a Polyhedral Oligomeric Silsesquioxane (POSS) Nanocomposite. Biomacromolecules, 7, 215-223.
14. Jha, M.K., Kumar, V., Maharaj, L. and Singh, R.J. (2004) Studies on Leaching and Recycling of Zinc from Rayon Waste Sludge. Industrial & Engineering Chemistry Research, 43, 1284-1295.
15. Sikorski, P., Wada, M., Heux, L., Shintani, H. and Stokke, B.T. (2004) Crystal Structure of Cellulose Triacetate I. Macromolecules, 37, 4547-4553.
16. Crompton, T.R. (1993) Practical Polymer Analysis. Plenum Press, New York.
17. Letcher, T.M. and Lutseke, N.S. (1990) A Closer Look at Cotton, Rayon, and Polyester Fibers. Journal of Chemical Education, 67, 361.
18. Menezes, E.A., Carapelli, R., Bianchi, S.R., Souza, S.N.P., Matos, W.O., Pereira-Filho, E.R. and Nogueira, A.R.A. (2010) Evaluation of the Mineral Profile of Textile Materials Using Inductively Coupled Plasma Optical Emission Spectrometry and Chemometrics. Journal of Hazardous Materials, 182, 325-330.
19. Cserháti, T. and Illés, Z. (1991) Comparison of Two Principal Component Analysis Methods to Evaluate Reversed- Phase Retention Data. Journal of Pharmaceutical and Biomedical Analysis, 9, 685-691.

CHAPTER 3

Chemometric Analysis of the Amino Acid Requirements of Antioxidant Food Protein Hydrolysates

Chibuike C. Udenigwe and Rotimi E. Aluko *

The Department of Human Nutritional Sciences and the Richardson Centre for Functional Foods and Nutraceuticals, University of Manitoba, Winnipeg, MB R3T 2N2, Canada

ABSTRACT

The contributions of individual amino acid residues or groups of amino acids to antioxidant activities of some food protein hydrolysates were investigated using partial least squares (PLS) regression method. PLS models were computed with amino acid composition and 3-z scale descriptors in the X-matrix and antioxidant activities of the samples in the Y-matrix; models were validated by cross-validation and permutation tests. Based on coefficients of the resulting models, it was observed that sulfur-containing (SCAA), acidic and hydrophobic amino acids had strong positive effects on scavenging of 2,2-diphenyl-1-picrylhydrazyl (DPPH) and H_2O_2 radicals in addition to ferric reducing antioxidant power. For superoxide radicals, only lysine and leucine showed strong positive contributions while SCAA had strong negative contributions to scavenging by the protein hydrolysates. In contrast, positively-charged amino acids strongly contributed negatively to ferric reducing antioxidant power and scavenging of DPPH and H_2O_2 radicals. Therefore, food protein hydrolysates containing appropriate amounts of amino acids with strong contribution properties could be potential candidates for use as potent antioxidant agents. We conclude that information presented in this work could support the development of low cost methods that will efficiently generate potent antioxidant peptide mixtures from food proteins without the need for costly peptide purification.

Keywords: Antioxidant properlerties; Free radicals; Reactive oxygen species; Chemometrics; Partial least square regression; Amino acids

1. INTRODUCTION

Antioxidant enzymatic food protein hydrolysates and food-derived peptides have gained particular interest as potential ingredients for formulation of functional foods and natural health products. Such formulated products could be used for human health sustenance especially for prevention and management of chronic diseases induced or propagated by oxidative stress. There is abundant information in the literature on food protein hydrolysates, peptide fractions and purified peptides with antioxidant activities in various oxidative models *in vitro* and in cell cultures. The antioxidant activity of food-derived peptides is based on scavenging of free radicals or reactive species [1,2], which is predominantly based on proton-coupled single electron or hydrogen atom transfer mechanisms [3,4]. In addition, food protein-derived antioxidant peptides also exhibit their activity by chelating pro-oxidant transition metals, e.g., Fe^{2+} and Cu^{2+}, reducing ferric ion, inhibiting oxidation of biological macromolecules such as unsaturated fatty acids [1,5] and cellular regulation of gene expression of antioxidant proteins, e.g., heme oxygenase-1 and ferritin [6]. Antioxidant activities of free amino acids have been evaluated [7] and a number of amino acids have been proposed to contribute positively to the antioxidant activity of purified food-derived and synthetic natural peptides. These amino acids include *Trp, Tyr, Met, Cys, His, Phe* and*Pro* [1,7–10]. Till date, several studies have attributed the antioxidative properties of many food protein hydrolysates, peptide fractions and purified peptides to these amino acids, but the chemistry and mechanisms of action have not been studied in detail. It is generally agreed upon that peptides possess substantially better antioxidant activities than their parent proteins and constituent amino acids [1], possibly due to increased accessibility of the functional side chain (R-group) to the reactive species, and the electron-dense peptide bonds. However, previous studies have indicated that some amino acids may be more active than their parent peptides [11,12].

For practical cost-effective application of food-derived peptides in the formulation of health-promoting food products, important limiting factors to be considered include a combination of high yields and potency of the natural bioactive ingredients. The expensive procedures and low product yield often associated with food protein-derived peptide purification and synthesis underscores the need to develop natural enzymatic food protein hydrolysates and peptide fractions that will possess potent antioxidant activities without the need for further extensive processing. In order to achieve this goal, initial directions of approach should involve elucidation of amino acid requirements for potency of food protein hydrolysates, and development of enzymatic hydrolysis and simple processing methods for enrichment of the desired amino acid residues in the peptide mixtures based upon their unique physicochemical properties. Indeed, the presence of the proposed antioxidant amino acid residues (*Trp, Tyr, Met, Cys, His, Phe and Pro*) may promote the antioxidant activities of food protein hydrolysates. However, the contributions of other amino acid residues to the oxidative system and the relative contribution/interactions of the so-called antioxidant amino acids in the complex peptide mixtures remain unclear. Moreover, the specific role of food-derived peptide antioxidants can be influenced by the type of oxidative species and chemistry of the reaction medium (e.g., pH) [2,3], which influence the properties of functional amino acid side chain groups. In-depth understanding of the relationship between composition and antioxidant activity of food protein hydrolysates and peptides is desirable for directed efforts towards the discovery of safer functional food ingredients [1]. Therefore, the objective of this study was to evaluate the contributions of individual amino acid residues, groups of amino acids (based on similar physicochemical properties of the R-group) and amino acid structural properties (based on 3-z scale amino acid descriptor) of food protein hydrolysates to antioxidant activities in four different oxidative assay systems using partial least squares (PLS) regression.

2. RESULTS AND DISCUSSIONS

PLS modeling is a widely used descriptive and predictive chemometrics approach for quantitative structure-activity relationship (QSAR) studies to elucidate how variation of molecular structures affect bioactivity of therapeutic agents, especially when working with high number of descriptor variables compared to the number of observations [13,14]. This method has been widely applied in food science research for developing models in QSAR studies of food-derived peptides such as bitter peptides [15,16], angiotensin converting enzyme inhibiting peptides [14,17–19], renin inhibiting peptides [20] and antimicrobial peptides [19]. Moreover, PLS modeling has been applied to the study of functional properties of food proteins and polypeptides; these are studies where the functional properties under investigation are impacted mostly by proportion of relevant amino acids rather than sequence [21]. Despite the wide application in food science, there is dearth of information in the literature on the use of chemometrics approaches, such as PLS method, in studying bioactivity of food protein hydrolysates that contain mixtures of peptides. Previous chemometric work on food-related bioactive compounds involved PLS analysis to elucidate the structural requirements for potency of synthesized antioxidant polyphenols in different chemical, cellular and enzymatic oxidative systems [22–24]; these methods often resulted in the discovery of compounds with more potent end-point antioxidative activities.

PLS modeling of amino acid parameters (X) and four different antioxidant activities (Y) of the food protein hydrolysates and fractions resulted in 12 models as shown in Table 1. The ferric reducing antioxidant property (FRAP) model (AA only) gave the best fit and predictive power whereas models for H_2O_2-scavenging displayed the lowest predictive powers (Figure 1 and Table 1). The R^2 values (Table 1) indicated that the PLS models explained 30% to 73% of the sum of squares in Y-variance for all the oxidative systems with up to 66% predictive ability (derived from Q^2_{cv}). The 12 models were theoretically validated initially by cross-validation during modeling and their predictive power also validated by permutation, where the bioactivity data were each randomly permuted a number of times but with unaltered X-variable followed by modeling of each permutation [13]. As shown in Table 1, repeated (20) rounds of permutation yielded cumulative R^2 (R^2_{cum}) intercept values of 0.006 to 0.314 and Q^2_{cv} intercept values of -0.223 to -0.008 for the 12 models, which are within suggested valid limits of R^2_{cum} intercept <0.4 and Q^2_{cv} intercept <0.05 for valid PLS models [25]. The t/u PLS score plots show relationships between X and Y variables in the AA + gAA + Σz_i (antioxidant activity + amino acid group + sum of z values) models for the four oxidative systems (Figure 1A–D). With the exception of the weak fit for H_2O_2, the data showed good fit and must have contributed positively to obtaining valid models that we used to determine relationships between amino acids or groups of amino acids and contributions to antioxidant potential of the food protein hydrolysates.

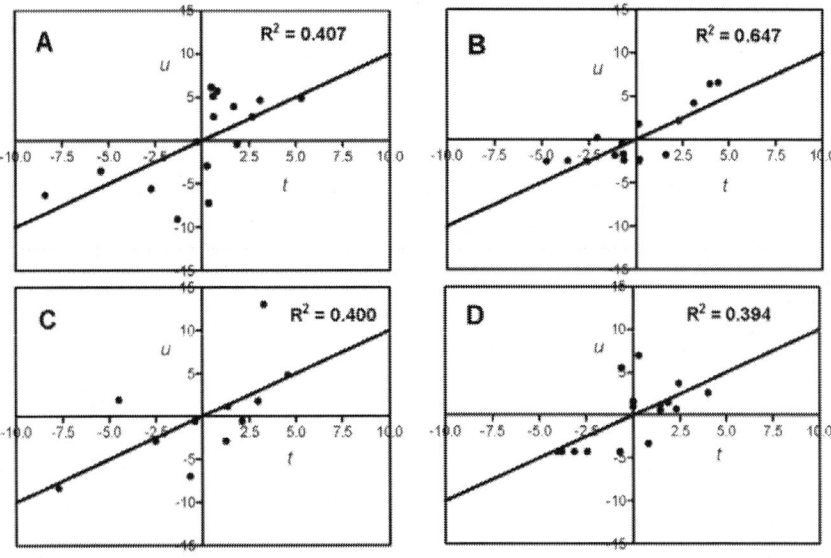

Figure 1. The t/u score plots of the partial least squares (PLS) models showing relationships between the antioxidant activity and amino acid descriptors (AA + gAA + Σz_i) of the food protein hydrolysates and peptide fractions; (**A**) DPPH radical scavenging; (**B**) ferric reducing antioxidant power (FRAP); (**C**) H_2O_2-scavenging, and (**D**) superoxide radical-scavenging.

Table 1. Summary of the partial least square regression models using amino acid compositions (AA), groups of amino acids (gAA) and sums of 3-z amino acid scales (Σz_i) of the food protein hydrolysates. The multiple correlation coefficient (R^2) estimates the model fit whereas the cross-validated correlation coefficient (Q^2_{cv}) indicates the models' predictive powers.

Antioxidant Property	N [a]	X [b]	A [c]	R^2 [d]	Q^2_{cv} [e]	Permutation Test [f] Int. R^2_{cum}	Int. Q^2_{cv}
DPPH-scavenging	16	AA + gAA + Σz_i	1	0.407	0.327	0.241	−0.096
		AA	1	0.448	0.358	0.288	−0.108
		gAA	1	0.405	0.231	0.149	−0.043
		Σz_i	1	0.304	0.288	0.006	−0.127
Ferric reducing	16	AA + gAA + Σz_i	1	0.647	0.604	0.166	−0.208
		AA	1	0.728	0.668	0.223	−0.219
		gAA	1	0.536	0.531	0.073	−0.169
H_2O_2-scavenging	11	AA + gAA + Σz_i	1	0.400	0.137	0.261	−0.122
		AA	1	0.467	0.136	0.314	−0.109
		Σz_i	1	0.340	0.232	0.088	−0.008
O_2^--scavenging	16	AA + gAA + Σz_i	1	0.394	0.179	0.212	−0.063
		gAA	2	0.602	0.142	0.166	−0.223

[a] N, number of observations used for PLS analysis; [b] X, X-variables (descriptors) in the validated PLS models; [c] A, number of significant components used in PLS modeling; [d] R^2, multiple correlation coefficients; [e] Q^2_{cv}, cross-validation correlation coefficients; [f] Permutation test, R^2_{cum} and Q^2_{cv} intercepts were calculated by SIMCA-P software during model validation.

The importance or relative contribution of the amino acid descriptor variables (X) in PLS modeling was obtained from the VIP plots (Figure 2A–D). The X variables with VIP > 1.0 are regarded as important with above average contribution while those with VIP < 0.5 are unimportant; variables with 1.0 > VIP > 0.5 could be important or not depending on the size of the dataset [26]. For example, the sulphur-containing (SC) amino acids as a group as well the individual amino acids that include *Met*, *Cys*, *Leu*, and *Lys* are strong contributors (VIP values > 1.0) to the superoxide model (Figure 2D). In contrast, *Gly*, *Ala*, *His*, *Asx*, and *Ile* are poor contributors to the superoxide model due to VIP values of <0.5. For our present study, X-variables were regarded as important strong contributors only when their VIP > 1.0 and weak contributors with VIP of 0.5–1.0. The relative contribution of X-variables in the PLS models depends on the value of the coefficients relative to the origin in the loading space [13]; in other words, the higher the coefficients in both directions, the more the contribution of the X-variable in explaining or predicting Y [27]. The coefficients (Figure 3A–D) and VIP plots (Figure 2A–D) indicated that the specific contribution of each or groups of amino acid and physicochemical properties (*3-z* scales) depends on the oxidative assay system. As shown in Figure 3A and Table 2, a high percentage composition of *Thr*, *Asx*, hydrophobic amino acids, as well as ↑Σz_3 (high electronic properties), ↓Σz_1 (low hydrophilicity or high hydrophobicity) and ↓Σz_2 (low bulk/molecular size) of the samples contributed positively to 2,2-diphenyl-1-picrylhydrazyl (DPPH)-scavenging. Interestingly, PCAA including *His* (highest VIP value) contributed negatively to the scavenging of DPPH. A similar pattern was also observed for H_2O_2-scavenging in addition to the strong positive contributions of *Cys*, *Phe*, *Leu* and *Pro* in this oxidative system (Figure 3C). Moreover, the SCAA (*Cys* + *Met*) were observed to be the strongest contributing amino acids for ferric reducing antioxidant power of the peptide mixtures whereas high amounts of *Lys* strongly reduced this activity (Figure 3B). In contrast, *Lys* and *Leu*composition supported superoxide radical scavenging by the samples while SCAA *Met* and *Cys* (highest VIP values) strongly reduced this activity. Interestingly, low z_1 (high hydrophobicity) character seem to have strong positive contributions to scavenging of the free radicals like DPPH and superoxide anion radical (O_2^-) as well as H_2O_2 (a reactive oxygen species) but not FRAP. The results agree with previous reports that indicate hydrophobic amino acids are able to interact better (when compared to hydrophilic amino acids) with the lipophilic environments that contain these free radicals. Table 2 also shows other amino acid compositions and properties that weakly contributed positively or negatively to the antioxidant activities of the peptide samples in the four oxidative systems.

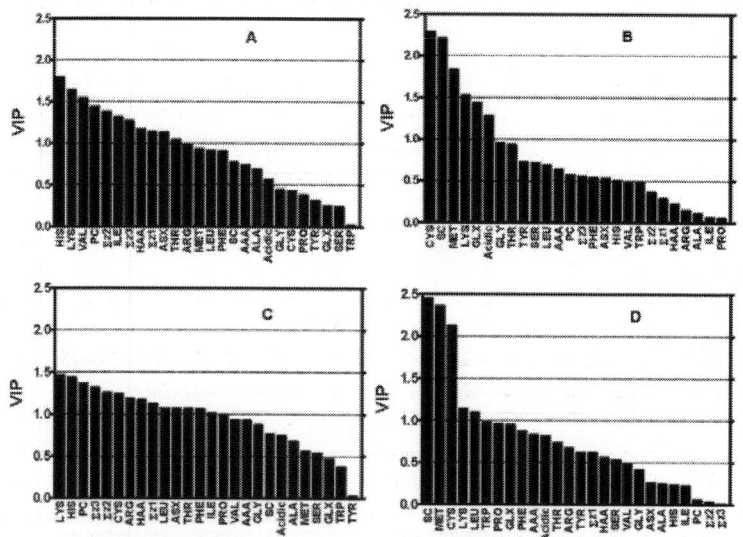

Figure 2. Variable Importance for the Projection (VIP) of the *3-z* scale models: (**A**) DPPH radical scavenging; (**B**) ferric reducing antioxidant power (FRAP); (**C**) H_2O_2-scavenging, and (**D**) superoxide radical-scavenging.

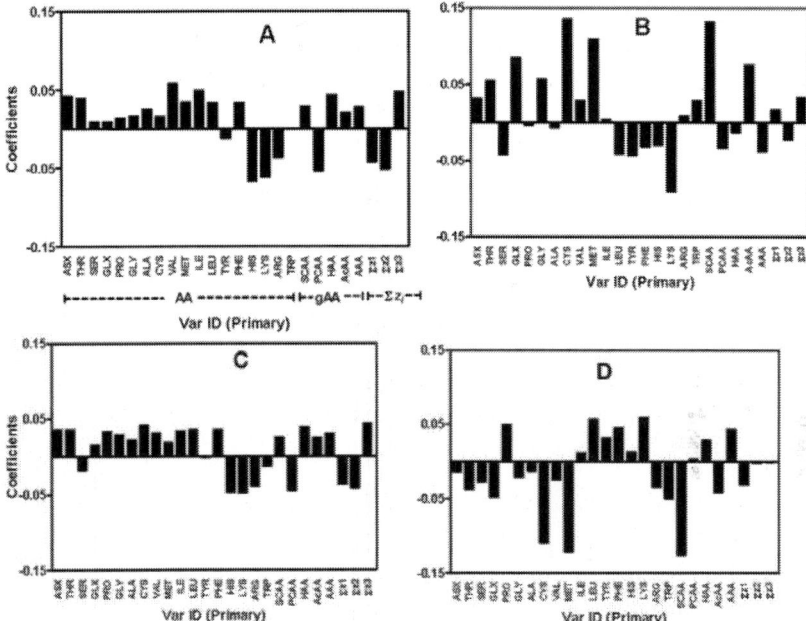

Figure 3. Coefficient plots of scaled and centered data of the partial least square regression models: (**A**) DPPH radical scavenging; (**B**) ferric reducing antioxidant power (FRAP); (**C**) H_2O_2-scavenging; and (**D**) superoxide radical-scavenging. The importance of a given X-variable is proportional to its distance (coefficient value) from the origin (zero). Above zero values indicate positive contributions while values less than zero indicate negative contributions.

Due to heterogeneity of reactive species implicated in human disease conditions, the structure and composition of amino acid residues required for potent antioxidant activity by food protein hydrolysates and peptides could depend on the type of reactive species and reaction conditions, e.g., pH and solvent. Consequently, high levels of particular amino acid residues can potentially increase or decrease the antioxidant activity of food protein hydrolysates depending on the reaction environment. Amino acid residues are major physiological targets of oxidants leading to the formation of stable and unstable oxidation products [28]; thus, the rate of transfer of the radicals to amino acids and stability of the reaction products can substantially influence potency of amino acids as antioxidants. Plausible mechanisms of radical scavenging or quenching activity of amino acid residues of proteins and peptides include proton or hydrogen atom donation by amino acids such as *Tyr*, *Trp* and *Cys* using their phenolic, indolic and sulfhydryl hydrogen, respectively [1,7,10]. The electron-dense side chain groups of*His*, *Trp* and *Met* also contribute to antioxidant properties of proteins and peptides [28]. In our study, the food protein hydrolysates in the dataset were composed of varying amounts of amino acid residues, which positively or negatively impacted their antioxidant activities based on the PLS models.

Table 2. Summary of contribution of amino acid compositions and properties to antioxidant activities of food protein hydrolysates based on partial least square regression coefficients and VIP plots of the AA + gAA + Σz_i models; strong contributors with VIP > 1.0 are shown in bold; (weak contributors with 1.0 > VIP > 0.5 are shown in brackets).

Oxidative System	Positive Contributors	Negative Contributors
DPPH radical scavenging	**Asp + Asn (Asx), Thr, Val, Ile, HAA** [a], **↑Σz_3** [b]**, ↓Σz_1** [c]**, ↓Σz_2** [d] (Ala, Met, Leu, Phe, SCAA [e], AcAA [f], AAA [g])	**His, Lys, Arg, PCAA** [h]
Ferric reducing	**Cys, Met, Glu + Gln (Glx), SCAA, AcAA** (Asp + Asp, Thr, Gly, ↑Σz_3)	**Lys** (Ser, Leu, Tyr, Phe, His, PCAA, AAA)
H_2O_2-scavenging	**Cys, Phe, Leu, Ile, Pro, Thr, Asx, HAA, ↑Σz_3, ↓Σz_1, ↓Σz_2** (Gly, Ala, Val, Met, SCAA, AcAA, AAA)	**His, Lys, Arg, PCAA** (Ser)
Superoxide radical scavenging	**Lys, Leu** (Pro, Phe, Tyr, HAA, ↓Σz_1)	**SCAA, Met, Cys** (Trp, Glx, AcAA, Thr, Arg, Ser)

[a] HAA, Hydrophobic amino acids (*Pro + Ala + Cys + Val + Met + Ile + Leu + Tyr + Phe + Trp*);
[b] Amino acids with ↑Σz_3 (electronic property) include *Cys > Pro > Asx > His > Trp*; [c] Amino acids with ↓Σz_1 (highly hydrophobic) include *Phe > Trp > Ile > Leu > Val > Met > Tyr*;
[d] Amino acids with ↓Σz_2 (lower side chain bulk/molecular size) include *Gly > Val > Thr > Ala > Ile > Ser > Leu > Cys*;
[e] SCAA, sulfur-containing amino acids (*Met + Cys*); [f] AcAA, Acidic amino acids (*Glx + Asx*); [g] AAA, Aromatic amino acids (*Phe + Tyr + Trp*);
[h] PCAA, positively charged amino acids (*Lys + Arg + His*).

The positive contributions of SCAA (*Cys* and *Met*) to the antioxidant activity of the samples (Table 2) can be attributed to their S-groups, which are prone to oxidation by the reactive species leading to the formation of stable oxidation products, cystine and methione sulfoxide, respectively [1,7,28]. The sulfhydryl group of *Cys* also acts as a strong reducing agent, hence its strong positive contribution in reducing ferric ion. Aspartic acid (*Asx*) and glutamic acid (*Glx*) can also donate their acidic hydrogen atoms near neutral pH, and this may have resulted to their strong contributions to the antioxidant activities of the samples. Similarly, *Phe* played a positive role in H_2O_2-scavenging and this can be attributed to susceptibility of its aromatic ring to oxidation. It is not clear how the HAA residues and low Σz_1 (low hydrophilicity or high hydrophobicity) of the samples directly contributed to antioxidant activity of food protein hydrolysates and fractions, although HAA may have interacted with DPPH via hydrophobic forces thereby increasing the proximity of the radical to the active functional groups. Moreover, amino acids with high Σz_3 exhibited positive effects in the assays except superoxide radical-scavenging, and this can be attributed to their electron density. Amino acids with less bulky structures (low Σz_2) correlated positively with antioxidant activities of the samples (Table 2), suggesting that steric effects can influence DPPH-scavenging activity of peptides, considering the bulky structure of the synthetic radical. On the other hand, the negative effects of the PCAA could be due to the fact that they can accept hydrogen ion near physiological pH and exist in the protonated state. Surprisingly, *His* residue was observed to negatively contribute to antioxidant properties of the peptide mixtures, based on the PLS models (Table 2). *His* can act as both hydrogen acceptor and donor near physiological pH; thus, its particular role in the complex peptide mixture could not be elucidated using these models. Moreover, the current work did not take into consideration the amino acid sequence of peptides, which could be important for *His*-containing peptides. Previously, the presence of *His* in synthetic peptides was reported to promote antioxidative activity in a linoleic acid oxidation model but not DPPH and superoxide radical-scavenging [8].

3. EXPERIMENTAL SECTION

3.1. X-Matrix

The *X*-matrix contained data for individual 18 amino acids AA (*Asx* for *Asn* + *Asp* and *Glx* for *Glu* + *Gln*) and groups of amino acids (gAA) such as sulphur-containing (SCAA-*Cys* and *Met*), positively charged (PCAA-*Arg*,*Lys* and *His*), acidic (AcAA-*Asx* and *Glx*), aromatic (AAA-*Phe*, *Tyr* and *Trp*) and hydrophobic amino acids (HAA-*Pro*, *Ala*, *Cys*, *Val*, *Met*, *Ile*, *Leu*, *Tyr*, *Phe* and *Trp*). The gAA were used because amino acids within each group have certain similarities (e.g., presence of sulphur or aromatic ring, *etc.*), which allow PLS modeling based on group characteristics. The results can be used to determine for example, whether aromatic rings contribute a greater to certain antioxidant property than acidic or sulphur groups. The *X*-matrix data for the amino acids of 16 samples were presented as percentage composition (Table 3) and the groups (Table 4) as sums of the respective data. The amino acid data were derived from previous studies that fractionated hempseed and pea protein hydrolysates based on molecular size [29], hydrophobic property [5] and net cationic property [30]. The amino acid 3-z scale was also used as *X*-variable in the multivariate descriptor matrix. The 3-z scales was previously derived by Hellberg *et al.* [31] by principal component analysis from 29 physicochemical variables of amino acids and were interpreted to be related to hydrophilicity (z_1), steric properties or side chain bulk/molecular size (z_2) and electronic properties (z_3) (see supplementary material). Algebraic sums of each of the 3-z scores (Σz_i) were calculated for each sample (see supplementary material) as previously reported [21] using Equation 1.

$$\sum z_i = \sum_{X=1}^{18} z_{iX} c_X \qquad (1)$$

Table 3. The % amino acid (AA) composition of the food protein hydrolysates used as variables in the X-matrix for partial least square regression analysis.

Sample ID	ASX[a]	THR	SER	GLX[b]	PRO	GLY	ALA	CYS	VAL	MET	ILE	LEU	TYR	PHE	HIS	LYS	ARG	TRP
1	11.39	3.68	4.63	20.06	4.00	4.29	4.47	1.32	4.66	1.81	3.84	6.75	3.45	4.60	2.78	2.97	14.07	1.23
2	9.49	3.60	4.73	15.18	3.19	3.23	4.91	0.29	5.67	1.94	4.15	9.91	4.78	7.68	2.61	3.19	13.87	1.58
3	11.70	3.77	4.79	19.31	4.04	3.93	4.77	0.66	5.26	2.03	4.16	7.26	3.50	5.01	2.47	2.94	12.96	1.44
4	12.79	4.01	4.69	22.71	4.23	4.54	4.30	1.26	4.45	1.85	3.98	5.15	3.06	3.21	2.47	2.56	13.60	1.16
5	12.70	4.00	4.47	22.87	4.89	4.71	4.12	1.58	4.24	1.80	3.90	4.82	3.62	2.85	2.49	2.51	13.31	1.11
6	13.79	3.60	6.20	13.92	5.15	3.76	5.01	0.24	5.63	0.91	5.43	9.91	3.87	7.41	1.61	6.10	6.83	0.68
7	13.94	3.89	6.63	17.12	2.33	3.52	5.54	0.18	5.23	0.70	4.13	8.70	2.77	3.97	2.49	9.07	9.79	0.00
8	10.63	3.86	5.71	14.78	6.47	5.00	4.30	0.39	4.45	1.70	4.04	6.68	5.33	7.76	3.28	7.35	8.00	0.27
9	12.59	3.34	6.19	13.75	5.14	3.96	5.03	0.39	4.13	0.87	6.71	9.95	7.15	8.73	1.90	4.26	5.15	0.74
10	10.85	3.11	4.41	12.87	5.42	4.66	3.44	0.38	5.82	1.07	5.85	14.57	5.09	12.03	1.81	3.31	3.97	1.36
11	11.04	3.22	3.82	6.64	8.05	3.26	3.62	0.29	7.68	0.68	9.13	19.48	2.44	16.44	0.63	1.20	1.22	1.16
12	15.09	4.43	5.16	20.30	5.16	4.12	5.47	0.41	4.77	1.07	4.99	10.16	3.41	6.81	1.05	3.40	3.12	1.06
13	11.23	3.42	3.96	22.14	5.78	3.83	4.31	0.26	4.28	0.88	3.23	6.55	4.12	4.50	3.55	8.03	9.32	0.63
14	9.86	3.59	5.73	10.81	4.56	3.47	4.62	0.13	3.64	0.96	3.21	8.22	3.21	6.20	3.76	16.38	10.19	1.46
15	8.12	2.68	3.72	9.47	2.37	3.87	4.18	0.12	2.66	0.51	2.25	8.92	5.61	4.86	4.22	11.79	24.50	0.14
16	6.61	1.74	4.24	9.04	2.93	2.00	2.48	0.11	2.02	0.21	2.14	6.64	3.48	2.24	4.26	16.76	30.18	2.93

[a] GLX = glutamine + glutamate; [b] ASX = asparagines + aspartate.

Table 4. Amino acid groups (gAA) present in food protein hydrolysates used as variables in the X-matrix for partial least square regression analysis.

Sample ID	SCAA[a]	PCAA[b]	HAA[c]	AcAA[d]	AAA[e]
1	3.13	19.82	36.13	31.45	9.28
2	2.22	19.67	44.10	24.67	14.05
3	2.70	18.37	38.14	31.00	9.95
4	3.11	18.62	32.64	35.50	7.42
5	3.39	18.31	32.95	35.56	7.58
6	1.15	14.54	44.24	27.71	11.96
7	0.88	21.35	33.55	31.06	6.74
8	2.09	18.63	41.39	25.41	13.36
9	1.26	11.31	48.84	26.34	16.62
10	1.45	9.09	55.03	23.72	18.48
11	0.97	3.05	68.97	17.68	20.04
12	1.48	7.58	43.32	35.39	11.28
13	1.14	20.90	34.53	33.37	9.24
14	1.09	30.33	36.21	20.67	10.88
15	0.63	40.51	31.63	17.59	10.62
16	0.33	51.19	25.18	15.65	8.65

[a] SCAA, sulphur-containing amino acids (Met + Cys); [b] PCAA, positively charged amino acids (Lys + Arg + His); [c] HAA, Hydrophobic amino acids; [d] AcAA, Acidic amino acids (Glx + Asx); [e] AAA, Aromatic amino acids (Phe + Tyr + Trp).

X represents each of the 18 amino acids in the protein hydrolysates and fractions, and c represents their percentage composition; the z values for Asx and Glx were calculated as averages of the z values of their respective constituent amino acids.

3.2. Y-Matrix

The Y-matrix contained the antioxidant data for the various observations, specifically the ability of the food protein hydrolysates to scavenge nitrogen-centered 2,2-diphenyl-1-picrylhydrazyl (DPPH) radical, superoxide anion radical (O_2^-) and H_2O_2, and their ferric reducing antioxidant power (FRAP). These antioxidant data were chosen because of the physiological relevance of the reactive species (O_2^- and H_2O_2) and reducing power in chronic human disease conditions, apart from DPPH-scavenging, which has been widely used in the literature as primary evaluation of antioxidant (reducing) capacity of food protein hydrolysates. The bioactivity data were presented as percentage scavenging of the free radicals and H_2O_2, and as absorbance at 700 nm for ferric reducing antioxidant power (see supplementary material). In order to ensure consistency in data interpretation, the antioxidant and amino acid composition data used in this study were reported by the same research group.

3.3. PLS Modeling

Modeling of the antioxidant activities (Y) as a function of the amino acid descriptors (X) of the protein hydrolysates was computed by the PLS method using SIMCA-P version 11.0 (Umetrics AB, Umeå, Sweden); the PLS models were generated for the four oxidative assay systems using individual (AA, gAA and Σz_i) and a combination of all the descriptors. All variables were centered and scaled to unit variance to ensure equal contribution in the models. The PLS models were validated theoretically using a combination of cross-validation and permutation tests [13]. The multiple correlation coefficient (R^2) and cross-validation correlation coefficient (Q^2_{cv}) were computed by SIMCA-P software and used to represent model fit and predictive ability, respectively. The relative contribution of the amino acid (X) descriptors to the antioxidant activities of the samples was computed by the software and presented as the Variable Importance for the Projection (VIP) and coefficient plots.

4. CONCLUSIONS

The present work has shown that chemometrics approach using PLS models successfully elucidated specific contributions (positive or negative) of individual and groups of amino acid residues to the antioxidative properties of food protein hydrolysates; however, the effects depend on the oxidative assay system. Based on the PLS models, it was observed that previously reported antioxidant amino acid residues (especially *His*) had negative influence on the antioxidant activities (DPPH, H_2O_2, superoxide, ferric reducing) studied in this work. Overall, low hydrophilic property (high hydrophobicity) was a strong positive contributor to scavenging of free radicals (but not ferric reducing ability) by food protein hydrolysates. In contrast to previous assumptions in the literature, aromatic amino acids did not show strong contributions to antioxidant systems studied in this work (except for H_2O_2). However, the data cannot be used to preclude strong contributions of aromatic amino acids to other antioxidant systems that were not included in our report. This is because antioxidant activities of food protein hydrolysates could also be influenced by the amino acid sequence and interactions between neighboring residues; these factors were not part of the PLS analysis in this work. Data from this work can serve as background for further chemometrics study on bioactive food protein hydrolysates using larger uniformly generated datasets and determining the effect of amino acid sequence. Results from this study could contribute to proper understanding of the structure-function relationships of antioxidant food protein hydrolysates and peptides. The results could also enhance development of enzymatic tools and processing conditions that will concentrate antioxidant amino acids into highly potent fractions. Future studies to apply the results of this study may include optimization of enzymatic hydrolysis and processing conditions to enrich the final peptide product with the desirable amino acid residues. The choice of amino acids or proportion of amino acids that need to be dominant in potent antioxidant food protein hydrolysates will depend on the target antioxidant system.

ACKNOWLEDGMENTS

This research work was supported by a Discovery grant to REA from the Natural Sciences and Engineering Research Council of Canada (NSERC).

REFERENCES

1. Elias, RJ; Kellerby, SS; Decker, EA. Antioxidant activity of proteins and peptides. *Crit. Rev. Food Sci. Nutr* **2008**, *48*, 430–441.
2. Udenigwe, CC; Lu, YL; Han, CH; Hou, WC; Aluko, RE. Flaxseed protein-derived peptide fractions: Antioxidant properties and inhibition of lipopolysaccharide-induced nitric oxide production in murine macrophages. *Food Chem* **2009**, *116*, 277–284.
3. Prior, RL; Wu, X; Schaich, K. Standardized methods for the determination of antioxidant capacity and phenolics in foods and dietary supplements. *J. Agric. Food Chem* **2005**, *53*, 4290–4302.

REFERENCES

4. Huang, D; Ou, B; Prior, RL. The chemistry behind antioxidant capacity assays. *J. Agric. Food Chem* **2005**, *53*, 1841–1856.
5. Pownall, TL; Udenigwe, CC; Aluko, RE. Amino acid composition and antioxidant properties of pea seed (*Pisum sativum* L.) enzymatic protein hydrolysate fractions. *J. Agric. Food Chem* **2010**, *58*, 4712–4718.
6. Erdmann, K; Grosser, N; Schipporeit, K; Schroder, H. The ACE inhibitory dipeptide *Met-Tyr* diminishes free radical formation in human endothelial cells via induction of heme oxygenase-1 and ferritin. *J. Nutr* **2006**, *136*, 2148–2152.
7. Hernández-Ledesma, B; Dávalos, A; Bartolomé, B; Amigo, L. Preparation of antioxidant enzymatic hydrolysates from -lactalbumin and -lactoglobulin. Identification of active peptides by HPLC-MS/MS. *J. Agric. Food Chem* **2005**, *53*, 588–593.
8. Chen, HM; Muramoto, K; Yamauchi, F; Fujimoto, K; Nokihara, K. Antioxidant properties of histidine-containing peptides designed from peptide fragments found in the digests of a soybean protein. *J. Agric. Food Chem* **1998**, *46*, 49–53.
9. Chen, HM; Muramoto, K; Yamauchi, F; Nokihara, K. Antioxidant activity of designed peptides based on the antioxidative peptide isolated from digests of a soybean protein. *J. Agric. Food Chem* **1996**, *44*, 2619–2623.
10. Saito, K; Jin, DH; Ogawa, T; Muramoto, K; Hatakeyama, E; Yasuhara, T; Nokihara, K. Antioxidative properties of tripeptide libraries prepared by the combinatorial chemistry. *J. Agric. Food Chem* **2003**, *51*, 3668–3674.
11. Erdmann, K; Cheung, BWY; Schroder, H. The possible roles of food-derived bioactive peptides in reducing the risk of cardiovascular disease. *J. Nutr. Biochem* **2008**, *19*, 643–654.
12. Kitts, DD; Weiler, K. Bioactive proteins and peptides from food sources. Applications of bioprocesses used in isolation and recovery. *Curr. Pharm. Des* **2003**, *9*, 1309–1323.
13. Wold, S; Sjöström, M; Eriksson, L. PLS-regression: a basic tool of chemometrics. *Chemom. Intell. Lab. Syst* **2001**, *58*, 109–130.
14. Pripp, AH; Isaksson, T; Stepaniak, L; Sørhaug, T; Ardö, Y. Quantitative structure-activity relationship modelling of peptides and proteins as a tool in food science. *Trends Food Sci. Technol* **2005**, *16*, 484–494.
15. Kim, HO; Li-Chan, ECY. Quantitative structure-activity relationship study of bitter peptides. *J. Agric. Food Chem* **2006**, *54*, 10102–10111.
16. Wu, J; Aluko, RE. Quantitative structure-activity relationship study of bitter di- and tri-peptides with angiotensin I-converting enzyme inhibitory activity. *J. Pept. Sci* **2007**, *13*, 63–69.

17. Wu, J; Aluko, RE; Nakai, S. Structural requirements of angiotensin I-converting enzyme inhibitory peptides: quantitative structure-activity relationship study of di- and tripeptides. *J. Agric. Food Chem* **2006**,*54*, 732–738.
18. Wu, J; Aluko, RE; Nakai, S. Structural requirements of angiotensin I-converting enzyme inhibitory peptides: quantitative structure-activity relationship modelling of peptides containing 4–10 amino acid residues. *QSAR Comb. Sci* **2006**, *25*, 873–880.
19. Yang, L; Shu, M; Ma, K; Mei, H; Jiang, Y; Li, Z. ST-scale as a novel amino acid descriptor and its application in QSAM of peptides and analogues. *Amino Acids* **2010**, *38*, 805–816.
20. Udenigwe, CC; Aluko, RE. Quantitative structure–activity relationship modeling of renin-inhibiting dipeptides. *Amino Acids* **2011**.
21. Siebert, KJ. Modeling protein functional properties from amino acid composition. *J. Agric. Food Chem* **2003**,*51*, 7792–7797.
22. Khalebnikov, AI; Schepetkin, IA; Domina, NG; Kirpotina, LN; Quinn, MT. Improved quantitative structure-activity relationship models to predict antioxidant activity of flavonoids in chemical, enzymatic, and cellular systems. *Bioorg. Med. Chem* **2007**, *15*, 1749–1770.
23. Om, A; Kim, JH. A quantitative structure-activity relationship model for radical scavenging activity of flavonoids. *J. Med. Food* **2008**, *11*, 29–37.
24. Roy, K; Mitra, I. Advances in quantitative structure-activity relationship models of antioxidants. *Expert Opin. Drug Discov* **2009**, *4*, 1157–1175.
25. Van der Voet, H. Comparing the predictive accuracy of models using a simple randomization test. *Chemom. Intell. Lab. Syst* **1994**, *25*, 313–323.
26. *SIMCA-P 11 Software Analysis Advisor*; version or edition; Umetrics AB: Umeå, Sweden, 2005.
27. Sandberg, M; Eriksson, L; Jonsson, J; Sjöström, M; Wold, S. New chemical descriptors relevant for the design of biologically active peptides. A multivariate characterization of 87 amino acids. *J. Med. Chem* **1998**,*41*, 2181–2491.
28. Davies, MJ. The oxidative environment and protein damage. *Biochim. Biophys. Acta* **2005**, *1703*, 93–109.
29. Girgih, AT; Udenigwe, CC; Aluko, RE. In vitro antioxidant properties of hempseed (*Cannabis sativa* L.) protein hydrolysate fractions. *J. Am. Oil Chem. Soc* **2011**, *88*, 381–389.
30. Pownall, TL; Udenigwe, CC; Aluko, RE. Effects of cationic property on the in vitro antioxidant properties of pea protein hydrolysate fractions. *Food Res. Int* **2011**, *44*, 1069–1074.

REFERENCES

31. Hellberg, S; Sjöström, M; Skagerberg, B; Wold, SJ. Peptide quantitative structure-activity relationships, a multivariate approach. *J. Med. Chem* **1987**, *30*, 1126–1135.

CHAPTER 4

QSAR Study of Antimicrobial 3-Hydroxypyridine-4-one and 3-Hydroxypyran-4-one Derivatives Using Different Chemometric Tools

Razieh Sabet and Afshin Fassihi *

Department of Medicinal Chemistry, Faculty of Pharmacy, Isfahan University of Medical Sciences, 81746-73461, Isfahan, Iran

ABSTRACT

A series of 3-hydroxypyridine-4-one and 3-hydroxypyran-4-one derivatives were subjected to quantitative structure-antimicrobial activity relationships (QSAR) analysis. A collection of chemometrics methods, including factor analysis-based multiple linear regression (FA-MLR), principal component regression (PCR) and partial least squares combined with genetic algorithm for variable selection (GA-PLS) were employed to make connections between structural parameters and antimicrobial activity. The results revealed the significant role of topological parameters in the antimicrobial activity of the studied compounds against *S. aureus* and *C. albicans*. The most significant QSAR model, obtained by GA-PLS, could explain and predict 96% and 91% of variances in the pIC_{50} data (compounds tested against *S. aureus*) and predict 91% and 87% of variances in the pIC_{50} data (compounds tested against *C. albicans*), respectively.

Keywords: 3-Hydroxypyridine-4-one; 3-hydroxypyran-4-one;QSAR; Chemometrics

1. INTRODUCTION

Quantitative structure activity relationships (QSAR) studies, as one of the most important areas in chemometrics, give information that is useful for molecular design and medicinal chemistry [1–5]. QSAR models are mathematical equations constructing a relationship between chemical structures and biological activities. These models have another ability, which is providing a deeper knowledge about the mechanism of biological activity. In the first step of a typical QSAR study one needs to find a set of molecular descriptors with the higher impact on the biological activity of interest [6–9]. A wide range of descriptors has been used in QSAR modeling. These descriptors have been classified into different categories, including constitutional, geometrical, topological, quantum chemical and so on. There are several variable selection methods including multiple linear regression (MLR), genetic algorithm (GA), partial least squares (PLS), principle component or factor analysis (PCA/FA), and so on. [7–9]. MLR yields models that are simpler and easier to interpret than PCR and PLS, because these methods perform regression on latent variables that don't have physical meaning. Due to the co-linearity problem in MLR analysis, one may remove the collinear descriptors before MLR model development. MLR equations can describe the structure activity relationships well but some information will be discarded in MLR analysis. On the other hand, factor analysis–based methods such as PLS regression can handle the collinear descriptors and therefore better predictive models will be obtained by PLS method [10].

It is almost 120 years since physicians revealed that the coincidence of blood and bacteria in a wound may cause a life-threatening infection. It has also been shown that blood or hemoglobin enhance the lethality of intraperitoneal or subcutaneous inocula of bacteria such as *Escherichia coli*. The effective component of hemoglobin is iron, and various soluble iron compounds exert an equivalent effect [11]. Administration of iron compounds to the host can increase the virulence of*Escherichia coli*, *Listeria monocytogenes*, *Salmonella typhimurium* and other pathogens [12]. In fact, iron is an essential element required for the growth and virulence of virtually all microbial pathogens [13, 14]. The availability of iron is critically important in host-parasite interactions [15]. Vertebrate hosts withhold iron from microbial invaders as a major defence mechanism against infection [13,15]. This task is achieved by sequestration of iron with iron-binding proteins, the most abundant, haemoproteins [16]. Some natural antibiotics, called siderophores, are low-molecular-weight chelating agents that form stable complexes with iron [17, 18]. There are many reports of the antimicrobial activity of chelating agents with different chemical structures [19–22]. Kojic acid (5-hydroxy-2-hydroxymethyl-pyran-4-one) and its 3-hydroxypyranones derivatives are examples of these compounds [19]. The bidentate chelating ligand 3-hydroxypyranone, which has a catechol-like function, forms stable complexes with several metal ions such as Fe^{3+}. *In vitro* antibacterial and antifungal activities of 3-hydroxy-pyridinones, bioisoster derivatives of 3-hydroxypyranones with metal chelating ability have been described. They have an inhibitory effect on the growth of*Escherichia coli, Listeria inocua* and *Staphylococcus aureus* [22]. More recently antibacterial and antifungal activities of carboxamide derivatives of 3-hydroxypyranones, 5-hydroxypyranones and 5-hydroxypyridinones have been reported [23, 24].

Few reports of antimicrobial studies of 3-hydroxypyridine-4-one and 3-hydroxypyran-4-one derivatives are available [19, 21–25] and in those they were not the subject of QSAR studies. Preliminary QSAR models for a series of such derivatives have been investigated by Fassihi *et al.*[25]. The antimicrobial activity against *C. albicans*, *S. aureus* and *P. aeroginosa* was the subject of MLR analysis in this preliminary study. MLR models revealed the best relationship between the antimicrobial activity and structural properties against *S. aureus* and *C. albicans*. In the present paper, more than 600 topological, geometrical, constitutional, functional group, electrostatic, quantum and chemical descriptors were used, for the development of QSAR equations, different methods were applied for the antimicrobial activity of the studied compounds against *S. aureus* and *C. albicans*. These methods where: (i) genetic algorithm - partial least squares (GA-PLS), (ii) MLR with factor analysis as the data pre-processing step for variable selection (FA-MLR) and (iii) principal component regression analysis (PCRA). The correlation coefficient (r), standard error of regression (SE), r^2cv (Q^2) and RMScv (STD(r)) were employed to judge the validity of regression equation.

2. EXPERIMENTAL SECTION

2.1. Software

The two-dimensional structures of molecules were drawn using the Hyperchem 7.0 software. The final geometries were obtained with the semi-empirical AM1 method in the Hyperchem program. The molecular structures were optimized using the Polak-Ribiere algorithm until the root mean square gradient was 0.01 kcal mol^{-1}. The resulted geometry was transferred into Dragon program package, which was developed by Milano Chemometrics and QSAR Group [26]. The z-matrix of the structures was provided by the software and transferred to the Gaussian 98 program. Complete geometry optimization was performed taking the most extended conformation as starting geometries. Semi-empirical molecular orbital calculation (AM1) of the structures was preformed using Gaussian 98 program [27]. MATLAB software (version 7.1 Math Work Inc.) was used for the PLS regression method.

2.2. Data Set And Descriptor Generation

The biological data used in this study are antimicrobial activity, (in terms of – log MIC), of a set of 3-hydroxypyridine-4-one and 3-hydroxypyran-4-one derivatives [23, 24, 25]. The structural features of these compounds are listed in Table 1 and then used for subsequent QSAR analysis as dependent variables. The large number of molecular descriptors was calculated using Hyperchem, Dragon package and Gaussian 98. Some chemical parameters including molecular volume (V), molecular surface area (SA), hydrophobicity (LogP), hydration energy (HE) and molecular polarizability (MP) were calculated using Hyperchem Software. Dragon software calculated different functional groups, topological, geometrical and constitutional descriptors for each molecule.

Table 1. Chemical structure of the compounds used in QSAR analysis.

Compound	X	R_2	R_3	R_5	R_6
1	NH	CH_3	OH	CH_2-R^a	H
2	NH	C_2H_5	OH	CH_2-R^a	H
3	NH	CH_3	OH	CH_2-$N(CH_3)_2$	H
4	NH	C_2H_5	OH	CH_2-$N(CH_3)_2$	H
5	NH	CH_3	OH	CH_2-$N(C_2H_5)_2$	H
6	NH	C_2H_5	OH	CH_2-$N(C_2H_5)_2$	H
7	N-Ph	CH_3	OH	H	H
8	N-m-OH-Ph	CH_3	OH	H	H
9	N-C_3H_7	CH_3	OH	H	H
10	N-C_4H_9	CH_3	OH	H	H
11	O	CH_2Cl	H	OH	H
12	O	CH_3	H	OH	H
13	O	CH_2OH	OH	H	CH_3
14	O	CH_2OH	OCH_2Ph	H	CH_3
15	O	CHO	OCH_2Ph	H	CH_3
16	O	COOH	OCH_2Ph	H	CH_3
17	O	$CONHR^b$	OCH_2Ph	H	CH_3
18	O	$CONHR^c$	OCH_2Ph	H	CH_3
19	O	$CONHR^d$	OCH_2Ph	H	CH_3
20	O	$CONHR^b$	OH	H	CH_3
21	O	$CONHR^c$	OH	H	CH_3
22	O	$CONHR^d$	OH	H	CH_3
23	O	CH_2OH	H	OCH_2Ph	H
24	O	COOH	H	OCH_2Ph	H
25	O	CONHPh	H	OCH_2Ph	H
26	N-CH_3	CONHPh	H	OCH_2Ph	H
27	N-CH_3	CONHPh	H	OH	H
28	O	CONH-R^e	H	OCH_2Ph	H
29	N-CH_3	CONH-R^e	H	OCH_2Ph	H
30	N-CH_3	CONH-R^e	H	OH	H
31	O	CH_2OH	H	OH	H

R^a is —N(piperidine), R^b is (pyridin-3-yl), R^c is (naphthalen-1-yl), R^d is (quinolin-3-yl), R^e is (4-methylcoumarin-7-yl)

2. EXPERIMENTAL SECTION

Gaussian 98 was employed for calculation of different quantum chemical descriptors including, dipole moment (DM), local charges, HOMO and LOMO energies. Hardness (η), softness (S), electronegativity (χ) and electrophilicity (ω) were calculated according to the method proposed by Thanikaivelan et al. [28].

Constitutional, topological, geometrical, functional group, quantum and physicochemical indices were used in this study; brief description of some of them is listed in Table 2.

Table 2. Brief description of some descriptors used in this study.

Constitutional	Mean atomic van der Waals volume (Mv) (scaled on Carbon atom), no. of heteroatoms, no. of multiple bonds (nBM), no. of rings, no. of circuits, no of H-bond donors, no of H-bond acceptors, no. of Nitrogen atoms (nN), chemical composition, sum of Kier-Hall electrotopological states (Ss), mean atomic polarizability (Mp), number of rotable bonds (RBN), mean atomic Sanderson electronegativity (Me), etc.
Topological	Narumi harmonic topological index (HNar), Total structure connectivity index (Xt), information content index (IC), mean information content on the distance degree equality (IDDE), total walk count, path/walk-Randic shape indices (PW3, PW4, PW5, Zagreb indices, Schultz indices, Balaban J index (such as MSD) Wiener indices, Information content index (neighborhood symmetry of 2-order) (IC2), Ratio of multiple path count to path counts (PCR), Lovasz-Pelikan index (leading eigenvalue) (LP1), total information content index (neighborhood symmetry of 1-order) (TIC1), reciprocal hyper-detour index (Rww), Average connectivity index chi-5 (X5A), piID (conventional bond-order ID number), etc.
Geometrical	3D Petijean shape index (PJI3), Asphericity (ASP), Gravitational index, Balaban index, Wiener index, Length-to-breadth ratio by WHIM (L/Bw), etc.
Quantum	Highest occupied Molecular Orbital Energy (HOMO), Lowest Unoccupied Molecular Orbital Energy (LUMO), Most positive charge (MPC), Sum of square of positive charges (SSPC), Sum of square of negative charges (SSNC), Sum of positive charges (SUMPC), Sum of negative charges (SUMNC), Sum of absolute of charges (SAC), Standard deviation (Std), Total dipole moment (DM_t), Molecular dipole moment at X-direction (DM_X), Molecular dipole moment at Y-direction (DM_Y), Molecular dipole moment at Z-direction (DM_Z), Electronegativity (χ= -0.5 (HOMO-LUMO)), Electrophilicity ($\omega = \chi^2/2\ \eta$), Hardness (η = 0.5 (HOMO+LUMO)), Softness (S=1/ η).
Functional group	Number of total secondary C(sp3) (nCs), Number of total tertiary carbons (nCt), Number of H-bond acceptor atoms (nHAcc), Number of secondary amides (aliphatic) (nCONHR), Number of unsubstituted aromatic C (nCaH), Number of ethers (aromatic) (nRORPh), Number of ketones (aliphatic) (nCO), Number of tertiary amines (aliphatic) (nNR2), Number of phenols (nOHPh), Number of total primary C(sp3) (nCp), etc.
Chemical	LogP (Octanol-water partition coefficient), Hydration Energy (HE), Polarizability (Pol), Molar refractivity (MR), Molecular volume (V), Molecular surface area (SA).

2.3. Data Screening and Model Building

The calculated descriptors were collected in a data matrix whose number of rows and columns were the number of molecules and descriptors, respectively. Genetic algorithm - partial least squares (GA-PLS), MLR with factor analysis as the data pre-processing step for variable selection (FA-MLR) and principal component regression analysis (PCRA) methods were used to derive the QSAR equations and feature selection was performed by the use of genetic algorithm (GA). The genetic algorithms are efficient methods for function minimization. In descriptor selection context, the prediction error of the model built upon a set of features is optimized [29].

In this study, to model the structure-antimicrobial activity relationships better, genetic algorithm-partial least square (GA-PLS) was employed [30, 31]. Partial least squares (PLS) linear regression is a recent technique that generalizes and combines features from principal component analysis and multiple regressions. PLS is a method suitable for overcoming the problems in MLR related to multicollinear or over-abundant descriptors [10].

Application of PLS method thus allows the construction of larger QSAR equations while still avoiding over-fitting and eliminating most variables. This method is normally used in combination with cross-validation to obtain the optimum number of components [32, 33]. The PLS regression method used was the NIPALS-based algorithm existed in the chemometrics toolbox of MATLAB software (version 7.1 Math Work Inc.). In order to obtain the optimum number of factors based on the Haaland and Thomas F-ratio criterion, leave-one-out cross-validation procedure was used [34].

In our previous study the classical approach of multiple regression technique was used for developing QSAR relation [25]. Here, FA-MLR was also performed on the dataset. Factor analysis (FA) was used to reduce the number of variables and to detect structure in the relationships between them. This data-processing step is applied to identify the important predictor variables and to avoid collinearities among them [35]. Principle component regression analysis, PCRA, was also tried for the dataset along with FA-MLR. With PCRA collinearities among X variables are not a disturbing factor and the number of variables included in the analysis may exceed the number of observations [36]. In this method, factor scores, as obtained from FA, are used as the predictor variables [35]. In PCRA, all descriptors are assumed to be important while the aim of factor analysis is to identify relevant descriptors.

3. RESULTS AND DISCUSSION

3.1. GA-PLS

In PLS analysis, the descriptors data matrix is decomposed to orthogonal matrices with an inner relationship between the dependent and independent variables. Therefore, unlike MLR analysis, the multicolinearity problem in the descriptors is omitted by PLS analysis. Because a minimal number of latent variables are used for modeling in PLS; this modeling method coincides with noisy data better than MLR. In order to find the more convenient set of descriptors in PLS modeling, genetic algorithm was used.

3. Results and Discussion

To do so, many different GA-PLS runs were conducted using different initial set of populations. The data set (compounds tested against *S. aureus*, n = 31) was divided into two groups: calibration set (n = 25) and prediction set (n = 6). Given 25 calibration samples; the leave-one-out cross-validation procedure was used to find the optimum number of latent variables for each PLS model. The most convenient GA-PLS model that resulted in the best fitness contained 17 indices, 5 of them being those obtained by MLR. The PLS estimate of coefficients for these descriptors are given in Figure 1.

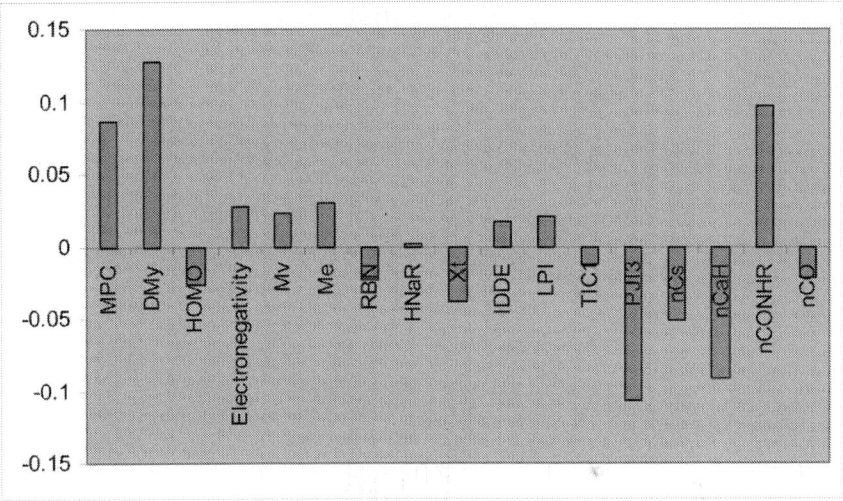

Figure 1. PLS regression coefficients for the variables used in GA-PLS model (against *S. aureus*).

As it is observed, a combination of quantum, topological, geometrical, constitutional, and functional group descriptors have been selected by GA-PLS to account the antimicrobial activity of the studied compounds. The majority of these descriptors are topological indices. The resulted GA-PLS model possessed very high statistical quality $R^2 = 0.96$ and $Q^2 = 0.91$. The values of pMIC using PLS model (refined from cross-validation or external prediction set) along with the corresponding relative errors of prediction (REP) are shown in Table 3. Very small values of relative errors confirm the accuracy of the proposed GA-PLS model for modeling antimicrobial activity of the studied compounds.

Table 3. Experimental and predicted activity of compounds against *Staphylococcus aureus*.

Compound	Experimental pMIC[a]	Predicted pMIC	REP[b] (%)
1	3.29	3.3205	0.9173
2	3.29	3.3007	0.3242
3	3.29	3.2266	-1.9664
4*	3.29	3.3976	3.1675
5	4.19	3.7498	-11.740
6	3.29	3.3205	0.9173
7	3.89	3.8255	-1.6850
8	3.29	3.2698	-0.6172
9	3.29	3.2886	-0.0440
10*	3.89	3.9283	0.9738
11	3.59	3.6207	0.8470
12	3.59	3.7254	3.6340
13	3.59	3.5063	-2.3883
14	3.59	3.6212	0.8627
15*	4.19	4.1563	-0.8119
16	3.59	3.5611	-0.8123
17	3.59	3.6177	0.7647
18	3.59	3.5548	-0.9915
19*	3.89	3.8950	0.1293
20	4.19	4.0995	-2.2079
21	3.59	3.7117	3.2787
22	5.10	5.0840	-0.3141
23	3.59	3.5533	-1.0318
24*	3.59	3.7223	3.5534
25	3.89	3.9222	0.8214
26	3.89	3.9779	2.2092
27	4.80	4.8022	0.0453
28	3.89	3.8591	-0.8011
29	3.59	3.4907	-2.8470
30*	4.49	4.5105	0.4549
31	3.59	3.4728	-3.3746

[a] pMIC= -log (MIC), [b] REP = Relative Error Prediction
*Compounds used as prediction set

The data set (compounds tested against *C. albicans*, n = 28) was again divided into two groups: calibration set (n = 23) and prediction set (n = 5). Given 23 calibration samples; the leave-one-out cross-validation procedure was used to find the optimum number of latent variables for each PLS model. Here, the most convenient GA-PLS model contained 15 indices, five of them being those obtained by MLR. The PLS estimate of coefficients for these descriptors are given in Figure 2. As it is observed, a combination of quantum, topological, geometrical and functional group descriptors have been selected by GA-PLS to account the antimicrobial activity of the compounds. The majority of these descriptors are topological indices again. The resulted GA-PLS model possessed very high statistical quality $R^2 = 0.91$ and $Q^2 = 0.87$. The values of pMIC using PLS model along with the corresponding REPs are shown in Table 4. Very small values of relative errors confirm the accuracy of the proposed GA-PLS model for modeling antimicrobial activity of the studied compounds.

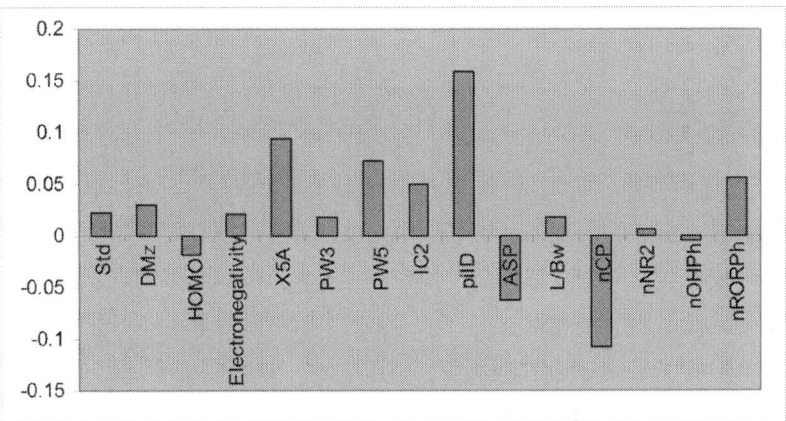

Figure 2. PLS regression coefficients for the variables used in GA-PLS model (against *C. albicans*).

3.2. FA-MLR and PCRA

Table 5 shows the five factor loadings of the variables (after VARIMAX rotation) for the compounds tested against *S. aureus*. As it is observed, about 79% of variances in the original data matrix could be explained by selected four factors.

Based on the procedure explained in the experimental section, the following three-parametric equation was derived.

pMIC = 4.786 (± 0.484) + 0.196 (± 0.063) DMy + 0.1666 (± 0.063) nCONHR – 0.130 (± 0.058) PJI3

$R^2 = 0.73$ S.E. = 0.31 F = 11.41 $Q^2 = 0.68$ RMScv = 0.34 N = 31 (1)

Equation 1 could explain 73% of the variance and predict 68% of the variance in pMIC data. This equation describes the effect of geometrical (PJI3), functional group (nCONHR) and quantum (DMy) indices on antimicrobial activity.

Table 4. Experimental and predicted activity of compounds against *Candida albicans*.

Compd.	Experimental pMIC	Predicted pMIC	REP(%)
2	3.29	3.4139	3.6304
4*	3.29	3.3893	2.9303
5	3.89	3.8920	0.0514
6	3.29	3.3591	2.0577
7	3.29	3.3835	2.7631
8	3.59	3.6477	1.5813
9	3.29	3.3208	0.9272
10*	3.59	3.6196	0.8175
11	3.89	3.9567	1.6857
12	3.89	3.7481	-3.7870
13	3.89	3.9092	0.4922
14	3.89	3.7076	-4.9191
15	3.89	3.8892	-0.0203
16	3.89	3.8422	-1.2433
17*	4.49	4.3961	-2.1360
18	4.49	4.4476	-0.9524
19	3.89	3.7076	-4.9191
20	3.89	3.8014	-2.3296
21	3.89	3.9525	1.5813
23	3.89	3.7450	-3.8727
24*	3.89	3.9056	0.3994
25	3.89	3.9969	2.6755
26	3.89	3.8489	-1.0691
27	3.89	3.7573	-3.5304
28	3.89	3.9503	1.5262
29*	3.89	3.9964	2.6619
30	3.89	3.8978	0.2006
31	3.89	3.8732	-0.4333

*Compounds used as prediction set

3. Results and Discussion

Table 5. Numerical values of factor loading numbers 1–4 for some descriptors after VARIMAX rotation (against *S. aureus*).

	1	2	3	4	Commonality
MPC	0.588	-0.105	0.587	-0.313	0.799
DMy	0.195	-0.054	0.762	0.071	0.627
HOMO	0.059	0.637	-0.013	0.620	0.794
Electonegativity	-0.643	-0.206	-0.199	-0.496	0.741
Mv	0.751	-0.413	0.362	-0.259	0.934
Me	0.001	-0.781	0.097	-0.298	0.708
RBN	0.087	0.902	0.068	0.003	0.826
HNar	0.866	0.051	0.217	-0.252	0.863
Xt	-0.645	-0.505	-0.307	0.081	0.772
IDDE	0.746	0.359	0.215	0.324	0.837
LP1	0.667	0.460	0.368	0.292	0.877
TIC1	0.714	0.413	0.175	0.127	0.726
PJI3	0.375	0.611	-0.315	-0.276	0.689
nCS	-0.559	0.578	-0.411	0.199	0.855
nCaH	0.894	-0.140	-0.143	-0.079	0.845
nCONHR	0.261	0.220	0.695	-0.906	0.765
nCO	-0.082	0.081	-0.214	0.853	0.787
pMIC *S. aureus*	0.041	-0.116	0.898	-0.051	0.824
%variance	29.87	20.10	17.15	12.12	79.24

When factor scores were used as the predictor parameters in a multiple regression equation using forward selection method (PCRA), the following equation was obtained:

$$pMIC = 3.756 \, (\pm 0.036) + 0.4000 \, (\pm 0.036) f_3$$
$$R^2 = 0.81 \quad S.E. = 0.19 \quad F = 35.05 \quad Q^2 = 0.79 \quad RMScv = 0.20 \quad N = 31 \quad (2)$$

Equation 2 also shows high equation statistics (81% explained variance and 79% predict variance in pMIC data). Since factor scores are used instead of selected descriptors, and any factor-score contains information from different descriptors, loss of information is thus avoided and the quality of PCRA equation is better than those derived from FA-MLR.

As it is observed from Table 5, in the case of each factor, the loading values for some descriptors are much higher than those of the others. These high values for each factor indicate that this factor contains higher information about which descriptors. It should be noted that all factors have information from all descriptors but the contribution of descriptor in different factors are not equal. For example, factors 1 and 2 have higher loadings for topological, constitutional and functional group indices, whereas information about quantum and functional group descriptors is highly incorporated in factors 3 and 4. Therefore, from the factor scores used by equation E_2, significance of the original variables for modeling the activity can be obtained. Factor score 1 indicates importance of Mv, HNar, nCaH and IDDE (topological, constitutional and functional group descriptors, respectively). Factor score 2 indicates importance of RBN and Me (constitutional descriptors), Factor score 3 and 4 signify the importance of DMy, and nCONHR (quantum and functional group descriptors, respectively).

Table 6 shows the five factor loadings of the variables (after VARIMAX rotation) for the compounds tested against *C. albicans*. As it is observed, about 80% of variances in the original data matrix could be explained by selected five factors.

Table 6. Numerical values of factor loading numbers 1–5 for some descriptors after VARIMAX rotation (against *C. albicans*).

	1	2	3	4	5	Commonality
Std	-0.491	-0.431	-0.459	-0.107	0.095	0.657
DMz	-0.007	0.102	-0.209	0.860	0.322	0.898
HOMO	0.240	0.811	-0.156	-0.349	0.014	0.861
Electonegativity	-0.706	-0.389	0.142	0.323	-0.310	0.871
X5A	-0.627	-0.664	-0.134	-0.102	0.129	0.879
PW3	-0.166	0.594	-0.377	-0.158	0.893	0.584
PW5	0.913	-0.079	0.055	0.135	-0.132	0.879
IC2	0.579	0.272	-0.164	0.210	0.584	0.820
piID	0.750	-0.070	-0.333	-0.190	-0.208	0.758
ASP	-0.075	0.087	0.866	-0.198	0.322	0.905
L/Bw	0.064	0.117	0.926	-0.023	0.164	0.902
nCp	-0.206	0.754	-0.224	-0.097	-0.325	0.777
nNR2	-0.366	0.722	0.148	0.287	-0.234	0.814
nOHPh	-0.191	-0.415	-0.165	-0.447	0.356	0.562
nRORPh	0.571	-0.522	0.379	0.002	-0.341	0.858
pMIC *C. albicans*	0.628	-0.627	-0.277	-0.107	0.602	0.872
%variance	22.58	20.58	14.71	14.02	8.71	80.60

Based on the procedure explained in the experimental section, the following four-parametric equation was derived.

$$pMIC = 5.980\ (\pm 0.695) + 0.182\ (\pm 0.022)\ piID - 0.167\ (\pm 0.024)\ nCp - 0.085\ (\pm 0.023)\ ASP - 0.058\ (\pm 0.023)\ PW3 \quad (3)$$

$R^2 = 0.81$ S.E. = 0.17 F = 34.76 $Q^2 = 0.79$ RMScv = 0.18 N = 28

3. Results and Discussion

Equation 3 could explain and predict 85% and 81% of the variance in pMIC data, respectively. This equation describes the effect of topological (piID and PW3), functional group (nCp) and geometrical (ASP) indices on the antimicrobial activity.

When factor scores were used as the predictor parameters in a multiple regression equation using forward selection method (PCRA), the following equation was obtained:

$$pMIC = 3.806\ (\pm 0.023) + 0.237\ (\pm 0.024) f_1 - 0.114\ ((\pm 0.024)\ f_3 + 0.081\ (\pm 0.024)\ f_2 - 0.065\ (\pm 0.024) f_4$$
$$R^2 = 0.83 \quad S.E. = 0.12 \quad F = 38.05 \quad Q^2 = 0.81 \quad RMScv = 0.12 \quad N = 28 \quad (4)$$

Equation 4 shows also high equation statistics (88% explained variance and 83% predicted variance in pMIC data). It should be noted that the variables (factor scores) used in Equation 4 are perfectly orthogonal to each other. Since factor scores are used instead of selected descriptors, and any factor-score contains information from different descriptors, loss of information is thus avoided and the quality of PCRA equation is better than those derived from FA-MLR.

As it is observed from Table 6, in the case of each factor, the loading values for some descriptors are much higher than those of the others. Factors 1 and 2 have higher loadings for topological, quantum and functional group indices, whereas information about geometrical, quantum and topological descriptors is highly incorporated in factors 3, 4 and 5. Therefore, from the factor scores used by equation E_4, significance of the original variables for modeling the activity can be obtained. Factor score 1 indicates importance of PW5, piID and electronegativity (topological and quantum descriptors). Factor score 2 indicates importance of HOMO nCp and nNR_2 (quantum and functional group descriptor). Factor score 3 signifies the importance of ASP and L/Bw (geometrical descriptors) and factor score 4 and 5 signify the importance of quantum and topological descriptors (DMz and PW3).

Comparison between the results obtained by GA-PLS and the other employed regression methods indicates higher accuracy of this method in describing antimicrobial activity of the studied compounds.

Difference in accuracy of the different regression methods used in this study is visualized inFigures 3 and 4 by plotting the predicted activity (by cross-validation) against the experimental values. Obviously, all linear models represented scattering of data around a straight line with slope and intercept close to one and zero, respectively. As it is observed, the plot of data resulted by GA-PLS represents the lowest scattering and that obtained by FA-MLR and PCR analysis have lower accuracy. It should be mentioned that the model which GA-PLS method provides is better than that MLR analysis provided in our previous study [25]. In fact, MLR analysis could explain and predict 55% and 35% of variances in the pMIC data (compounds tested against *S. aureus*) and predict 82% and 73% of variances in the pMIC data (compounds tested against *C. albicans*).

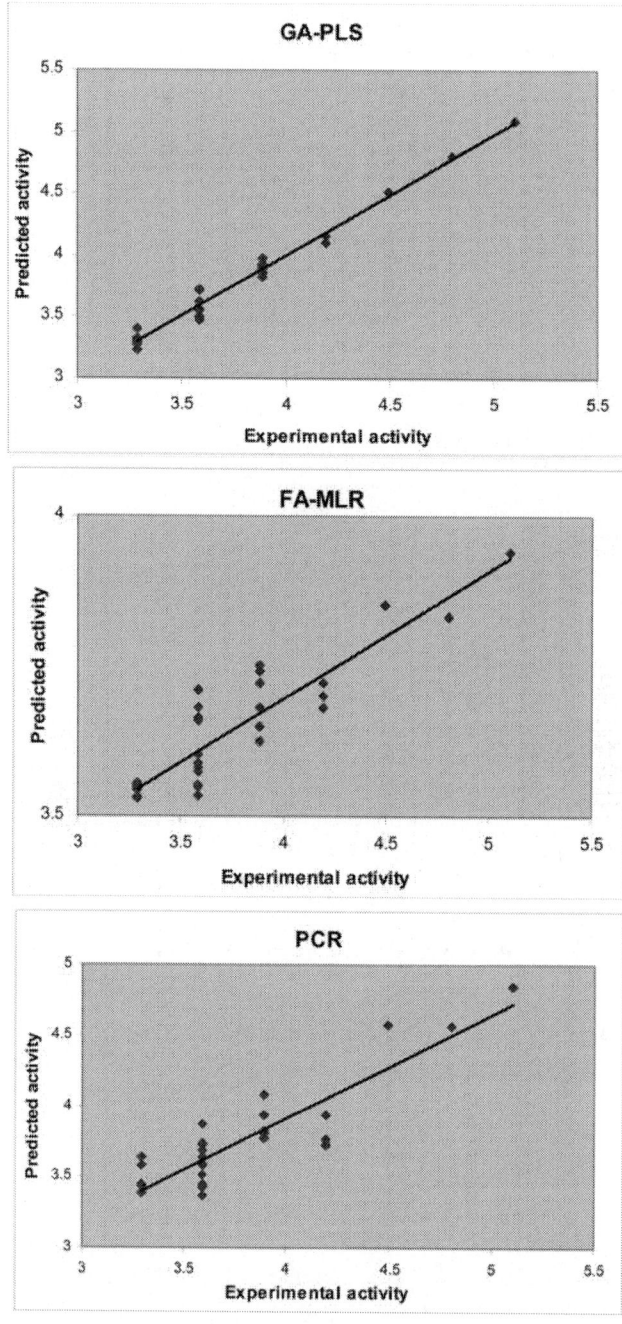

Figure 3. Plots of the cross-validated predicted activity against the experimental activity for the QSAR models obtained by different chemometrics methods (against *S. aureus*).

3. Results and Discussion

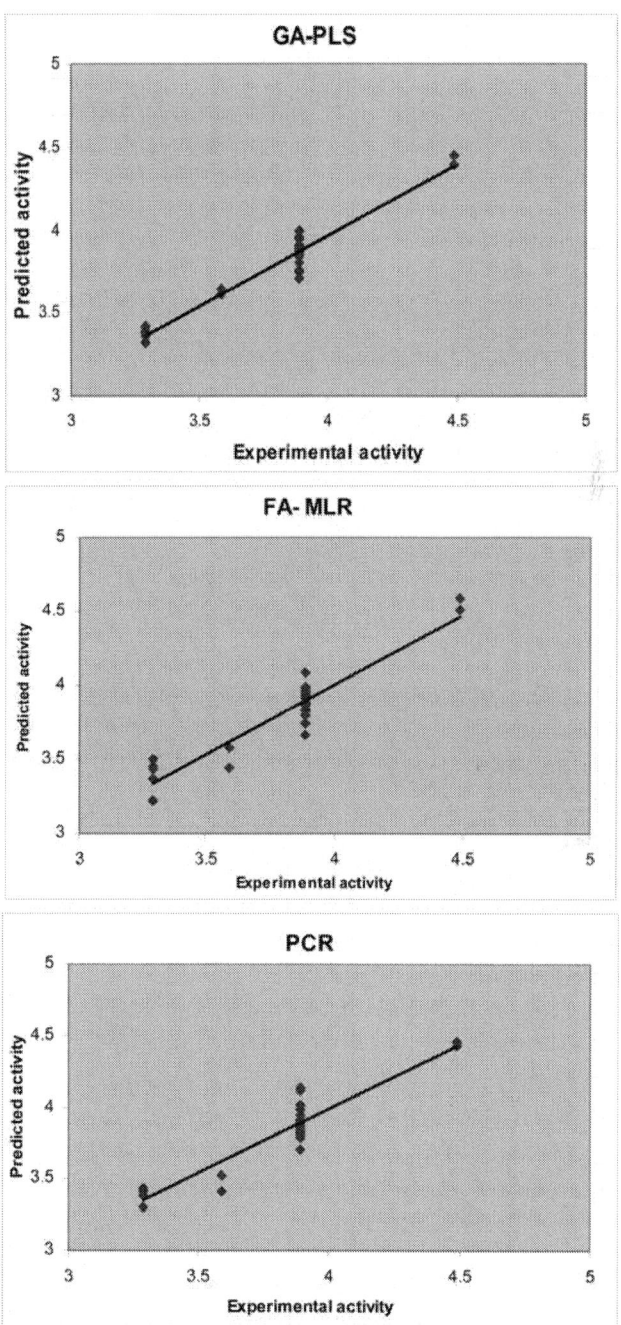

Figure 4. Plots of the cross-validated predicted activity against the experimental activity for the QSAR models obtained by different chemometrics methods (against *C. albicans*).

4. CONCLUSIONS

Quantitative relationships between molecular structure and inhibitory activity of a series of 3-hydroxypyridine-4-one and 3-hydroxypyran-4-one derivatives were discovered by a collection of chemometrics methods including GA-PLS, FA-MLR and PCRA. The results revealed the significant role of topological parameters in the antimicrobial activity of the studied compounds against *S. aureus* and *C. albicans*. A comparison between the different statistical methods employed indicated that GA-PLS represented superior results and it could explain and predict 96% and 91% of variances in the pMIC data (compounds tested against *S. aureus*) and predict 91% and 87% of variances in the pMIC data (compounds tested against *C. albicans*). As it is observed, the plot of data resulted by GA-PLS represents the lowest scattering, and the impact of topological descriptors was the most.

ACKNOWLEDGMENTS

This work was supported by Isfahan Pharmaceutical Sciences Research Center.

REFERENCES

1. Schmidi, H. Multivariate Prediction for QSAR. *Chemom. Intell. Lab. Sys* **1997**, *37*, 125–134.
2. Hansch, C; Kurup, A; Garg, R; Gao, H. Chem-bioinformatics and QSAR. A Review of QSAR Lacking Positive Hydrophobic Terms. *Chem. Rev* **2001**, *101*, 619–672.
3. Wold, S; Trygg, J; Berglund, A; Antii, H. Some Recent Developments in the PLS Modeling.*Chemom. Intell. Lab. Syst* **2001**, *58*, 131–150.
4. Hemmateenejad, B; Miri, R; Akhond, M; Shamsipur, M. QSAR Study of the Calcium Channel Antagonist Activity of some Recently Synthesized Dihydropyridine Derivatives. An Application of Genetic Algorithm for Variable Selection in MLR and PLS Methods.*Chemom. Intell. Lab. Syst* **2002**, *64*, 91–99.
5. Hemmateenejad, B; Miri, R; Akhond, M; Shamsipur, M. Quantitative Structure Activity Relationship Study of Recently Synthesized 1,4-Dihydropyridine Calcium Channel Antagonists. Application of Hansch Analysis Methods. *Arch. Pharm. Pharm. Med. Chem* **2002**,*10*, 472–480.
6. Horvath, D; Mao, B. Neighborhood Behavior. Fuzzy Molecular Descriptors and Their Influence on the Relationship between Structural Similarity and Property Similarity. *QSAR. Comb. Sci* **2003**, *22*, 498–509.

REFERENCES

7. Putta, S; Eksterowicz, J; Lemmen, C; Stanton, R. A Novel Subshape Molecular Descriptor. *J. Chem. Inf. Comput. Sci* **2003**, *43*, 1623–1635.
8. Gupta, S; Singh, M; Madan, AK. Superpendentic Index: A Novel Topological Descriptor for Predicting Biological Activity. *J. Chem. Inf. Comput. Sci* **1999**, *39*, 272–277.
9. Consonni, V; Todeschini, R; Pavan, M. Structure/Response Correlations and Similarity/Diversity Analysis by GETAWAY Descriptors. 2. Application of the Novel 3D Molecular Descriptors to QSAR/QSPR Studies. *J. Chem. Inf. Comput. Sci* **2002**, *42*, 693–705.
10. Deeb, O; Hemmateenejad, B; Jaber, A; Garduno-Juarez, R; Miri, R. Effects of the Electronic and Physicochemical Parameters on the Carcinogenecis Activity of Some Sulfa Drug Using QSAR Analysis Based on Genetic-MLR & Genetic-PLS. *Chemosphere* **2007**, *67*, 2122–2130.
11. Eaton, JW; Brandt, P; Mahoney, JR; Lee, JT, Jr. Haptoglobin: A Natural Bacteriostat. *Science* **1982**, *215*, 691–693.
12. Jones, RL; Peterson, CM; Grady, RW; Kumbaraci, T; Cerami, A. Effects of Iron Chelators and Iron Overload on *Salmonella* Infection. *Nature* **1977**, *267*, 63–65.
13. Weinberg, ED. Cellular Iron Metabolism in Health and Diseased. *Drug Metab. Rev* **1990**, *22*, 531–579.
14. Weinberg, ED. Iron and Infection. *Microbiol. Rev* **1978**, *42*, 45–66.
15. Weinberg, ED. Iron Withholding: A Defense Against Infection and Neoplasia. *Physiol. Rev* **1984**, *64*, 65–102.
16. Skaar, EP; Gaspar, AH; Schneewind, O. *Bacillus anthracis* IsdG, a Heme-Degrading Monooxygenase. *J. Bacteriol* **2006**, *188*, 1071–1080.
17. Neilands, JB. Siderophores: Structure and Function of Microbial Iron Transport Compounds. *J. Biol. Chem* **1995**, *270*, 26723–26726.
18. Sebat, JL; Paszczynski, AJ; Cortese, MS; Crawford, RL. Antimicrobial Properties of Pyridine-2,6-Dithiocarboxylic Acid, a Metal Chelator Produced by *Pseudomonas spp*. *Appl. Envir. Microbiol* **2001**, *67*, 3934–3942.
19. Weinberg, GA. Iron Chelators as Therapeutic Agents against *Pneumocystis carinii*. *Antimicrob.Agents Chemother* **1994**, *38*, 997–1003.
20. van Asbeck, BS; Marcelis, JH; Marx, JJM; Struyvenberg, A; van Kats, JH; Verhoef, J. Inhibition of Bacterial Multiplication by the Iron Chelator Deferoxamine: Potentiating Effect of Ascorbic Acid. *Eur. J. Clin. Microbiol. Infect. Dis* **1983**, *2*, 426–431.
21. Erol, DD; Yulug, N. Synthesis and Antimicrobial Investigation of Thiazolinoalkyl-4(1H)-pyridones. *Eur. J. Med. Chem* **1994**, *29*, 893–897.

22. Min-Hua, F; van der Does, L; Bantjes, A. Iron (III)-Chelating Resins. 3. Synthesis, Iron (III)-Chelating Properties, and *in vitro* Antibacterial Activity of Compounds Containing 3-hydroxy-2-methyl-4(1H)-pyridinone Ligands. *J. Med. Chem* **1993**, *36*, 2822–2827.
23. Aytemir, MD; Erol, DD; Hider, RC; Ozalp, M. Synthesis and Evaluation of Antimicrobial Activity of New 3-Hydroxy-6-methyl-4-oxo-4*H*-pyran-2-carboxamide Derivatives. *Turk. J. Chem* **2003**, *27*, 757–764.
24. Aytemir, MD; Hider, RC; Erol, DD; Ozalp, M; Ekizoglu, M. Synthesis of New Antimicrobial Agents; Amide Derivatives of Pyranones and Pyridinones. *Turk. J. Chem* **2003**, *27*, 445–452.
25. Fassihi, A; Abedi, D; Saghaie, L; Sabet, R; Fazeli, H; Bostaki, Gh; Deilami, O; Sadinpour, H. Synthesis, Antimicrobial Evaluation and QSAR Study of Some 3-hydroxypyridine-4-one and 3-hydroxypyran-4-one Derivatives. In *Eur. J. Med. Chem*; 2008; DOI:10.1016/j.emech.2008.10.022.
26. Todeschini, R. Milano Chemometrics and QSPR Group, http://michem.disat.unimib.it/, accessed 9 September, 2008.
27. Frisch, MJ; Trucks, MJ; Schlegel, HB; Scuseria, GE; Robb, MA; Cheeseman, JR; Zakrzewski, VG; Montgomery, JA; Stratmann, JR; Burant, JC; *et al. Gaussian 98, Revision A.7*; Gaussian, Inc: Pittsburgh PA, 1998.
28. Roy, K. QSAR of Adenosine Receptor Antagonists II: Exploring Physicochemical Requirements for Selective Binding of 2-arylpyrazolo [3,4-c] quinoline Derivatives with Adenosine A^1 and A^3 Receptor Subtypes. *QSAR. Comb. Sci* **2003**, *22*, 614–621.
29. Siedlecki, W; Sklansky, J. On Automatic Feature Selection. *Int. J. Pattern Recog. Artif. Intell* **1988**, *2*, 197–220.
30. Leardi, R. Application of Genetic Algorithm-PLS for Feature Selection in Spectral Data Sets. *J. Chemomtr* **2000**, *14*, 643–655.
31. Leardi, R; Gonzalez, AL. Genetic Algorithm Applied to Feature Selection in PLS Regression: How and When to Use Them. *Chemom. Intell. Lab. Syst* **1998**, *41*, 195–207.
32. Fassihi, A; Sabet, R. QSAR Study of p56[lck] Protein Tyrosine Kinase Inhibitory Activity of Flavonoid Derivatives Using MLR and GA-PLS. *Int. J. Mol. Sci* **2008**, *9*, 1876–1892.
33. Leardi, R. Genetic Algorithms in Chemometrics and Chemistry: A Review. *J. Chemometrics* **2001**, *15*, 559–569.
34. Hemmateenejad, B. Optimal QSAR Analysis of the Carcinogenic Activity of Drugs by Correlation Ranking and Genetic Algorithm-Based. *J. Chemometrics* **2004**, *18*, 475–485.

35. Franke, R; Gruska, A. Chemometrics Methods in Molecular Design. In *Methods and Principles in Medicinal Chemistry*; van Waterbeemd, H, Ed.; VCH: Weinheim, Germany, 1995; Volume 2, pp. 113–119.
36. Kubinyi, H. The Quantitative Analysis of Structure-Activity Relationships. In *Burger's Medicinal Chemistry and Drug Discovery*, 5th Ed.; Wolff, ME, Ed.; Wiley: New York, USA, 1995; Volume 1, pp. 506–509.

CHAPTER 5

Comprehensive and Comparative Metabolomic Profiling of Wheat, Barley, Oat and Rye Using Gas Chromatography-Mass Spectrometry and Advanced Chemometrics

Bekzod Khakimov*, Birthe Møller Jespersen and Søren Balling Engelsen

Department of Food Science, Faculty of Science, University of Copenhagen, Rolighedsvej 30, Frederiksberg C, 1958 Copenhagen, Denmark

ABSTRACT

Beyond the main bulk components of cereals such as the polysaccharides and proteins, lower concentration secondary metabolites largely contribute to the nutritional value. This paper outlines a comprehensive protocol for GC-MS metabolomic profiling of phenolics and organic acids in grains, the performance of which is demonstrated through a comparison of the metabolite profiles of the main northern European cereal crops: wheat, barley, oat and rye. Phenolics and organic acids were extracted using acidic hydrolysis, trimethylsilylated using a new method based on trimethylsilyl cyanide and analyzed by GC-MS. In order to extract pure metabolite peaks, the raw chromatographic data were processed by a multi-way decomposition method, Parallel Factor Analysis 2. This approach lead to the semi-quantitative detection of a total of 247 analytes, out of which 89 were identified based on RI and EI-MS library match. The cereal metabolome included 32 phenolics, 30 organic acids, 10 fatty acids, 11 carbohydrates and 6 sterols. The metabolome of the four cereals were compared in detail, including low concentration phenolics and organic acids. Rye and oat displayed higher total concentration of phenolic acids, but ferulic, caffeic and sinapinic acids and their esters were found to be the main phenolics in all four cereals. Compared to the previously reported methods, the outlined protocol provided an efficient and high throughput analysis of the cereal metabolome and the acidic hydrolysis improved the detection of conjugated phenolics.

Keywords: GC-MS; Metabolomics; Barley; Wheat; Oat; Rye; TMSCN; PARAFAC2

1. INTRODUCTION

Cereals such as wheat, barley, rye and oat are amongst the mostly grown agricultural food products worldwide and the most important cereal crops for human consumption in northern Europe. The detailed chemical and functional composition of these crops is defining their use for food and feed as well as their prices. Cereals are the most important study objects in foodomics studies seeking to optimize their health beneficial factors and/or reducing deleterious metabolites. While the gross chemical composition, such as carbohydrates, proteins, dietary fibers and micronutrient contents, are important characteristics of cereal products, recent studies showed that relatively low concentration secondary metabolites such as antioxidant phenolics, organic acids and phytosterols have a significant influence on the health and nutritional values of cereals [1,2]. The beneficial health effects associated with the consumption of cereals have been attributed to dietary fiber content [3] as well as phenolics that possess antioxidant, radical scavenging and cholesterol lowering properties [4,5,6,7]. Whole grain barley intake has proven to decrease the low-density lipoprotein (LDL) cholesterol in an intervention study involving hypercholesterolemic patients [8]. Moreover, phenolic acids were found to be important texturizing agents in cooking-extrusion of cereals [9] and recognized as the main antioxidant constituents of cereals [10].

Quantitative and qualitative analysis of both, secondary and primary metabolites (with molecular weight of up to 1500 Da) of grains are studied within cereal metabolomics. Cereal metabolomics offers an insight into the metabolic fluctuations of cereal cultivars that may reveal effects of genetic modifications as well as of biotic and abiotic stresses [11]. Recent studies have illustrated the power of cereal metabolomics to reveal effects of growth temperature [12], salt stress [13], drought stress [14], and biotic stress [15]. Cereal metabolomics is also a promising approach to reveal biochemical and genetic backgrounds of quality traits and may open new possibilities towards targeted breeding [16,17].

Comprehensive metabolomic profiling of cereals requires a reliable protocol that enables extraction of maximum metabolic information in a high-throughput and reproducible manner. Metabolomics studies performed for uncovering single and/or multiple internal and/or external effects on cereals aim to cover as broad range of metabolites as possible. However, due to the great physico-chemical diversity of cereal metabolites, it is in practice impossible to cover the whole cereal metabolome using a single protocol. The phytochemical composition, including phenolics of wheat [18,19,20], barley [21], oat [22] and rye [23] have been investigated in a number of studies within the HEALTHGRAIN diversity-screening program [24].

This study demonstrates the development of comprehensive GC-MS metabolomics protocol for profiling a broad range of phenolics and organic acids from whole grain flour samples, and applied on wheat, barley, rye and oat. Phenolics of cereals are primarily present in conjugated and bonded forms with carbohydrates, lipids and other cell membrane components that alter their solubility and thus bioavailability [21]. Analysis of phenolic content of cereals is mainly performed by basic hydrolysis of cereal extracts [18], which can only cleave ester bonds and stabilize de-esterification reactions. However, a substantial part of phenolics and other organic acids of cereals are conjugated through glycosidic and/or ether bonds to carbohydrates and other molecules. In contrast to basic hydrolysis, acidic hydrolysis allows the cleavage of not only ester bonds, but also glycosidic and ether bonds at an elevated temperature. The advantages of this approach have been demonstrated in polyphenol analysis of the wheat and rice grains [25,26].

In this study, a standardized, high-throughput and unbiased protocol was developed for GC-MS metabolomic profiling of free and conjugated phenolics and organic acids of whole-grain cereals using hydrochloric acid based hydrolysis followed by trimethylsilyl derivatization. The study demonstrates the first application of a novel trimethylsilylation method based on trimethylsilyl cyanide (TMSCN) for derivatization of cereal metabolites. When compared to other frequently used derivatization methods, the new protocol provides a more unbiased and broad-spectrum derivatization of metabolites and is able to provide reproducible metabolomics profiles of complex biological samples [27]. The obtained raw GC-MS data of cereals were processed by a semi-automated multi-way decomposition method, PARAFAC2 [28]. The PARAFAC2 processing of the raw GC-MS data lead to unambiguous deconvolution of elusive peaks such as, overlapped, retention time shifted and low s/n peaks and enable an automatic estimation of relative concentrations of detected peaks [29,30]. Metabolite extraction and GC-MS analysis of the cereal samples were performed within a bigger study, which involved a larger set of barley samples (manuscript in preparation). The main aim of this study was to demonstrate the performance of the protocol, using new technologies within metabolomics, and to show first results of a comparative application to the four major north European cereals: wheat, barley, rye and oat. To the best of our knowledge, this is the first study illustrating a comprehensive GC-MS profiling of phenolics and organic acids of cereals using exactly the same protocol across different cereals.

2. EXPERIMENTAL SECTION

Whole grain samples of wheat (*Tr. aestivum*, variety Bussard), barley (*H. vulgare*, variety Bomi), rye (*S. cereal*, variety Petkus) and oat (*A. sativa*, variety Sang) were purchased in Sepetember 2012 from the Danish bread cereal producing company Aurion (Hjørring, Denmark). All four cereals were grown under biodynamical conditions in Jutland during the season 2011/12.

2.1. Metabolite Extraction and Sample Derivatization

Cereal metabolites were extracted from 50 mg of milled grains that were soaked into 600 µL 85% methanol and vortexed for 20 s at 3000 rpm followed by 20 min incubation at 30 °C using a Thermomixer (Model 5436, Eppendorf, Hamburg, Germany) at 1400 rpm. After 3 min of centrifugation at 16,000× g, the supernatant was transferred to a fresh 2 mL Eppendorf tube (Hamburg, Germany) and the remaining flour sample was extracted a second time using the same extraction procedure. Then, the combined extracts were completely dried under nitrogen gas flow at 40 °C and hydrolyzed by using 240 µL of 6 M hydrochloric acid at 96 °C for 1 h by stirring at 1400 rpm. The hydrolyzed extracts were transferred into a fresh 2 mL glass vials and phenolics and organic acids were extracted into diethyl ether. Ether-based extraction of phenolics and organic acids was performed twice, by addition of 800 µL diethyl ether and vortexing for 25 s. The obtained ether fractions were completely dried using nitrogen gas flow and re-solubilized in 200 µL 100% methanol. Aliquots, 90 microliter, of the final extracts were transferred into 200 µL glass inserts and completely dried under nitrogen gas flow, sealed and stored at −20 °C until GC-MS analysis. Each sample was spiked with an internal standard (IS) (5 µL of 0.2 mg mL^{-1} solution of ribitol). In order to avoid any moisture, the samples stored in the freezer were dried under reduced pressure before derivatization. Sample derivatization and injection were fully automated by using a Multi-Purpose Sampler (MPS, GERSTEL, Mülheim, Germany) with DualRait WorkStation integrated to a GC-MS system from Agilent (CA, USA). Each sample was individually derivatized by addition of 40 µL trimethylsilyl cyanide (TMSCN) and incubated for 40 min at 40 °C. Two replicate samples per cereal were analyzed in randomized order and the MPS autosampler allowed a sequential derivatization of all samples in the same manner by keeping the derivatization time constant, throughout the analysis.

2.2. GC-MS Data Acquisition

The GC-MS consisted of an Agilent 7890A GC and an Agilent 5975C series MSD. GC separation was performed on a Phenomenex ZB 5MSi column (30 m × 250 µm × 0.25 µm). A derivatized sample volume of 1 µL was injected into a cooled injection system (CIS port) using Solvent Vent mode at the vent pressure of 7 kPa until 0.3 min after injection at the vent flow of 100 mL min^{-1}. Detailed information on CIS and MPS parameters are described in Khakimov *et al.* 2013 [27]. Hydrogen was used as carrier gas, at a constant flow rate of 1.2 mL min^{-1}, and the initial temperature of CIS was set to 120 °C for 0.3 min followed by heating at 5 °C s^{-1} until reaching 320 °C and then held for 10 min. The GC oven program was as follows: initial temperature 40 °C, equilibration time 3.0 min, heating rate 12.0 °C min^{-1}, end temperature 300 °C, hold time 8.0 min and post run time 5 min at 40 °C. Mass spectra were recorded in the range of 50–500 *m/z* with a scanning frequency of 3.2 scans s^{-1}, and the MS detector was switched off during the 8.5 min of solvent delay time and after 25.5 min of the run time. The transfer line, ion source and quadrupole temperatures were set to 290, 230 and 150 °C, respectively.

The mass spectrometer was tuned according to manufacturer's recommendation by using perfluorotributylamine (PFTBA).

2.3. Data Analysis

Initial analysis and visualization of the GC-MS data was performed using ChemStation software (Agilent, Germany). Retention indices of detected metabolites were calculated using the Van den Dool and Kratz equation and retention times of C10-C40 alkanes that were analyzed using the same GC-MS protocol [31]. The raw GC-MS data was imported from netCDF format to .mat files into Matlab® ver. R2012b (8.0.0.783) and data was manually divided into 121 smaller baseline separated intervals in retention time dimension. Each interval was modeled separately by PARAFAC2 as described previously [30]. PARAFAC2 modeled the three-way raw GC-MS data (elution time × mass spectra × samples) without any prior data pre-processing. The PARAFAC2 model outcomes: the elution profiles, which represent the TIC in the raw data, and spectral profiles, which represent the experimental EI-MS of deconvoluted peaks, were used for metabolite identification. The PARAFAC2 resolved mass spectrum of each peak was extracted and compared against NIST05 library (NIST, USA), Golm Metabolite Database [32]. Finally, PARAFAC2 concentration profiles, which represented relative concentrations of detected peaks were extracted and normalized according to the peak area of the internal standard (ribitol). The obtained metabolite table was used for exploring variations of phenolics in cereals and for principal component analysis (PCA) [33] after autoscaling of the data.

3. RESULTS AND DISCUSSION

3.1. GC-MS Metabolomic Profiling and PARAFAC2 Based Data Processing

The total ion current (TIC) chromatograms of the GC-MS data obtained from hydrolyzed extracts of the four cereals are illustrated in Figure 1. Just over 300 peaks with a s/n ratio >10 were detected from GC-MS profiles. Validated PARAFAC2 models of 121 intervals of the raw GC-MS data revealed 389 components including resolved peaks, shoulders of neighbor peaks and baseline. Then, each PARAFAC2 model was individually evaluated and components that represent baseline, artifact peaks such as column bleed and reagent derived peaks and shoulders of neighbor peaks were eliminated, resulting in 247 chromatographic peaks with unique retention indices and mass spectra. The PARAFAC2 modeling of GC-MS intervals representing vanillin, protocatechuic acid and β-resorcylic acid are demonstrated in Figure 2.

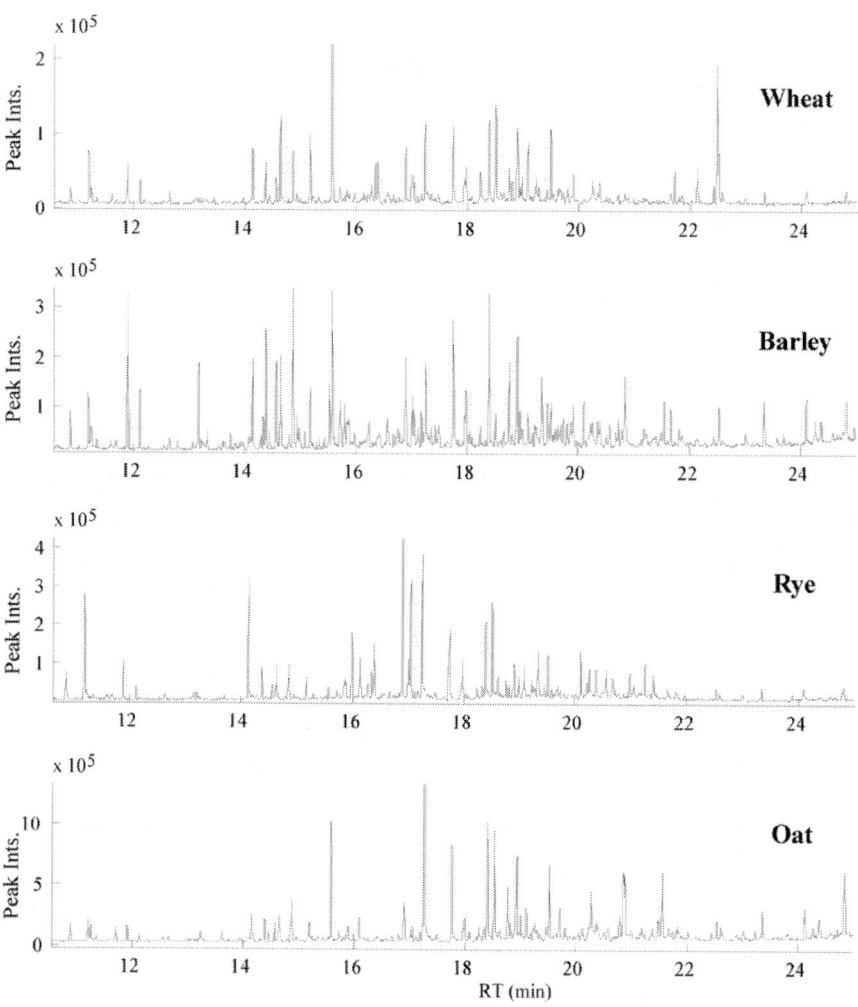

Figure 1. The total ion current (TIC) chromatograms of GC-MS data obtained on wheat, barley, rye and oat metabolite extracts.

Comparison of RIs and PARAFAC2 resolved mass spectra of 247 resolved peaks against the NIST05 and Golm Metabolite Database resulted in the identification of 89 metabolites (Table 1) at level 2 as described in Metabolomics Standards Initiative report [34]. A total of 32 out of 89 identified metabolites were trimethylsilyl (TMS) derivatives of phenolic acids, their esters and aldehydes. In addition to the previously found phenolic acids from different barley genotypes [21], several other phenolics such as p-salicylic, gallic, gentisic, homovanillic and α-resorcylic acids and methyl esters of ferulic, caffeic, protocatechuic and sinapinic acids were identified. Small molecular organic acids, alcohols and their esters constituted 30 out of

89 identified metabolites. These included succinic, glyceric, maleic, fumaric, malic, pyroglutamic, azelaic acids and methyl esters of aconitic and citric acids that are part of the same or different metabolic pathways, and in addition, TMS-derivatives of 10 fatty acids and their esters, 6 sterols and a flavonoid, catechin-nTMS.

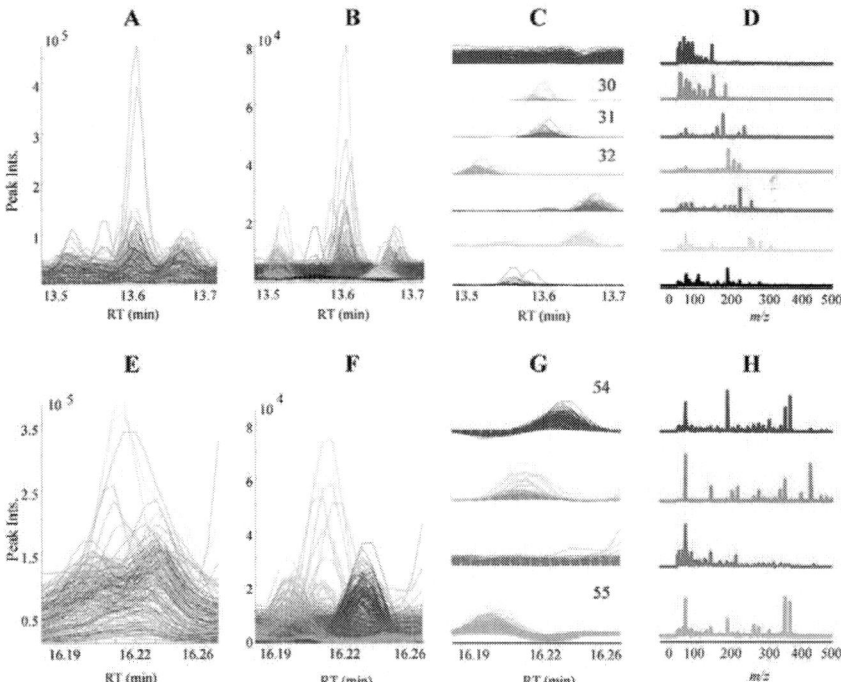

Figure 2. PARAFAC2 based processing of raw GC-MS data intervals. (**A**) and (**E**) are the TIC of raw GC-MS data intervals. (**B**) and (**F**) are the superimposed PARAFAC2 elution profiles of the raw GC-MS data intervals with seven and four components, respectively. (**C**) and (**G**) are subplots of (**B**) and (**F**), respectively. * Numbers of elution profiles correspond to the metabolites represented in Table 1. (**D**) and (**H**) are subplots of PARAFAC2 mass spectral profiles.

Table 1. A list of identified metabolites from wheat, barley, rye and oat flour samples by GC-MS. Metabolite identification was performed at level 2 as described in Metabolomics Standards Initiative report [34] and was based on RI and EI-MS library match (>80). [a] Metabolites with more than one isomers and/or TMS-derivatives; [b] tentatively identified.

No	Metabolites	RT min	RI (r)	RI (c)
1.	Laevulic acid-1TMS	9.04	1030	1070
2.	Sorbic acid-1TMS	9.06	1009	1071
3.	Hepta-2,4-dienoic acid, methyl ester	9.28	1000	1080
4.	Octanol-1-1TMS	9.51	1101	1090
5.	Malonic acid-2TMS	9.99	1205	1207
6.	(3,3-Dimethyl-1-cyclohexen-1-yl)oxy]-1TMS	9.97	1110	1206
7.	Benzoic acid-1TMS	10.42	1228	1226
8.	3-Methyl-2-furoic acid-1TMS	10.38	1107	1224
9.	Glycerol-3TMS	10.88	1282	1246
10.	1,3-Dihydroxypropanone-2-2TMS	11.03		1249
11.	Succinic acid-2TMS	11.24	1292	1262
12.	Glyceric acid-3TMS	11.51	1199	1274
13.	Maleic acid-2TMS	11.55	1286	1275
14.	Fumaric acid-2TMS	11.60	1178	1278
15.	*p*-Hydroxybenzaldehyde-1TMS	11.85	1280	1289
16.	2-Hydroxyheptanoic acid-2TMS	11.83	1312	1288
17.	3-Hydroxybutanoic acid-2TMS	12.12	1403	1401
18.	Resorcinol-2TMS	12.2	1378	1404
19.	Trimethyl aconitate	12.50	1428	1419
20.	Citric acid, trimethyl ester	12.82	1442	1435
21.	3-Hydroxyanthranilic acid, methyl ester-1TMS	12.8		1434
22.	2,4-Dihydroxy-5-methylpyrimidine-2TMS	12.89	1403	1439
23.	5-Hydroxy-2-(hydroxymethyl)-4H-pyran-4-one-2TMS	13.08	1492	1448
24.	Maseptol-1TMS	13.12	1358	1450
25.	Malic acid-2TMS	13.19	1494	1453
26.	2-Hydroxycyclohexanecarboxylic acid-2TMS	13.23	1402	1456
27.	3-Hydroxyoctanoic acid-2TMS	13.35	1452	1462
28.	Pyroglutamic acid-2TMS	13.46	1466	1467
29.	Erythritol-4TMS	13.47		1467
30.	Dimethyl azelate	13.61	1485	1474
31.	4-Hydroxybenzeneacetic acid, methyl ester-1TMS	13.62	1458	1475
32.	Vanillin-1TMS	13.55	1469	1471
33.	Citric acid, trimethyl ester-1TMS	13.76		1482
34.	2-Furancarboxylic acid, 5-[(oxy)methyl]-1TMS	13.72	1540	1480
35.	4-Hydroxyphenylethanol-2TMS	13.92	1475	1490
36.	Anozol	14.15	1603	1601

37.	2-Ketoglutaric acid-3TMS	14.34	1622	1612
38.	3-Methyl-3-hydroxypentanedioic acid-3TMS	14.3	1610	1609
39.	Dodecane-6-hydroxy-1TMS	14.40	1631	1615
40.	4-Hydroxybenzoic acid-2TMS	14.45	1618	1618
41.	Methyl Isovanillate-1TMS	14.66	1547	1629
42.	Suberic acid-2TMS	15.11	1682	1654
43.	Syringaldehyde-1TMS	15.15	1658	1656
44.	β-D-Arabinopyranose-4TMS [a]	15.23	1692	1660
45.	β-D-Xylopyranose-4TMS	15.30	1694	1664
46.	3,5-Dihydroxybenzoic ac. met.est.-2TMS	15.35	1656	1667
47.	2,5-Dimethoxymandelic acid-2TMS	15.38	1867	1669
48.	Vanillic acid-2TMS	15.72	1656	1687
49.	4-Hydroxycinnamic acid, methyl ester-1TMS	15.88	1565	1696
50.	Azelaic acid-2TMS	15.98	1800	1802
51.	2,3-Dihydroxyphosphoric acid, propyl ester-4TMS	15.86	1708	1695
52.	Methyl 2-(oxy)-2-(4-(oxy)phenyl)propanoate-2TMS	16.14	1757	1811
53.	α-D-Galactofuranoside, methyl-2,3,5,6-tetrakis-4TMS [a]	16.11	1845	1810
54.	3,5-Dihydroxy benzoic ac.-3TMS	16.24	1826	1818
55.	3,4-Dihydroxy benzoic ac.-3TMS	16.20	1826	1815
56.	D-Fructose-5TMS	16.41	1867	1828
57.	Isocitric acid-4TMS	16.34	1835	1823
58.	Catechin-nTMS [a]	16.44		1830
59.	Homovanilic acid-2TMS	16.4	1867	1827
60.	β-D-Galactopyranoside, methyl 2,3,4,6-tetrakis-4TMS [a]	16.68	1900	1844
61.	Catechin-nTMS [a]	16.77		1849
62.	2,5-Dihydroxy benzoic ac.-3TMS	16.78	1796	1850
63.	α-D-Glucopyranoside, methyl 2,3,4,6-tetrakis-4TMS [a]	16.90	1928	1857
64.	Syringic acid-2TMS	16.88	1845	1856
65.	β-D-Glucopyranoside, methyl 2,3,4,6-tetrakis-4TMS [a]	17.05	1928	1866
66.	α-D-Glucopyranose, 1,2,3,4,6-pentakis-5TMS [a]	17.02	1924	1864
67.	Palmitic acid, methyl ester	17.01	1870	1864
68.	D-Galactose, 2,3,4,5,6-pentakis-5TMS [a]	17.12	1970	1871
69.	p-Coumaric acid-2TMS	17.18	1924	1874
70.	Ferulic acid, methyl ester-1TMS	17.25	1765	1878
71.	3,4,5-Trihydrozy benzoic ac.-4TMS	17.45	1976	1890
72.	2-Hydroxymandelic acid, ethyl ester-2TMS	17.34	1777	1884
73.	4'-Cyclohexylacetophenone	17.58	1703	1898
74.	Caffeic acid methyl ester-2TMS	17.76	1863	2010
75.	β-D-Glucopyranose-5TMS [a]	17.75	1970	2009

76.	2-Hydroxysebacic acid-3TMS	18.13	2059	2034
77.	Ferulic acid-2TMS	18.40	2076	2052
78.	8,11-Octadecadienoic acid, methyl ester	18.35	2093	2049
79.	Sinapinic acid methyl ester-1TMS	18.51	1943	2059
80.	Methyl vanillactate-2TMS	18.55	2030	2062
81.	Caffeic acid-3TMS	18.76	2114	2076
82.	9-Methoxy-4α-methyl-2,3,7-trihydroxy-4,4a-dihydro-2H-benzo[c]chromen-6(3H)-one [b]	18.85		2082
83.	Linoleic acid-1TMS	19.23	2202	2207
84.	4,8-Dihydroxy-2-quinolinecarboxylic acid-3TMS	19.46	2265	2224
85.	Sinapinic acid-2TMS	19.52	2221	2228
86.	Androsterone type plant sterol [b]	19.89		2254
87.	3-Hydroxyandrostan-17-one-1TMS	19.98	2186	2261
88.	19-Norandrosterone-3-TMS [b]	20.36	2198	2288
89.	9,10-Dihydroxystearic acid-3TMS	20.87	2517	2426
90.	3,7-di-Hydroxy-androstan-17-one-2TMS	21.09	2432	2443
91.	9,10-Dihydroxystearic acid, dimethyl ester-2TMS	21.49	2784	2474
92.	2,3-Dihydroxypalmitic acid, propyl ester-2TMS	21.84	2581	2601
93.	2-Deoxy-6-phosphogluconolactone-5TMS	23.26		2820
94.	2-Hydroxytetracosanoic acid, methyl ester-1TMS	23.69	2894	2858
95.	3,7-Dihydroxycholest-5-ene-2TMS	23.95	2900	2881

3.2. Principal Component Analysis (PCA)

In order to explore the metabolomics data, PCA was performed on the metabolite table, including eight cereal samples in duplicates and 89 identified metabolites. PC1 *versus* PC2 scores plot of the PCA model (Figure 3A) show a clear separation of four different cereals explaining more than 60% variation of the data. The loadings plot of the corresponding model (Figure 3B) demonstrates a large spread of the 89 metabolites and revealed no clear groupings of metabolites classes. However, major part of the benzoic acid derived phenolics such as 3,5-dihydroxybenzoic, 3,4-dihydroxybenzoic and 3,4,5-trihydroxybenzoic acids are grouped on the upper left part of the loadings plot showing greater abundance in barley compared to the other cereals. In contrast to this, cinnamic acid derived phenolics such as ferulic, sinapinic and syringic acids are located on the bottom right corner showing greater concentrations in rye and wheat. Phenolics such as caffeic and 4-hydroxybenzoic acids have the highest concentrations in oat and significantly contribute to its separation from other cereals. However, detailed variations of phenolics and organic acids within and between cereal cultivars require a closer investigation of the data. In the following section, univariate comparisons of some metabolites are represented and the findings are compared to previous results reported in the literature.

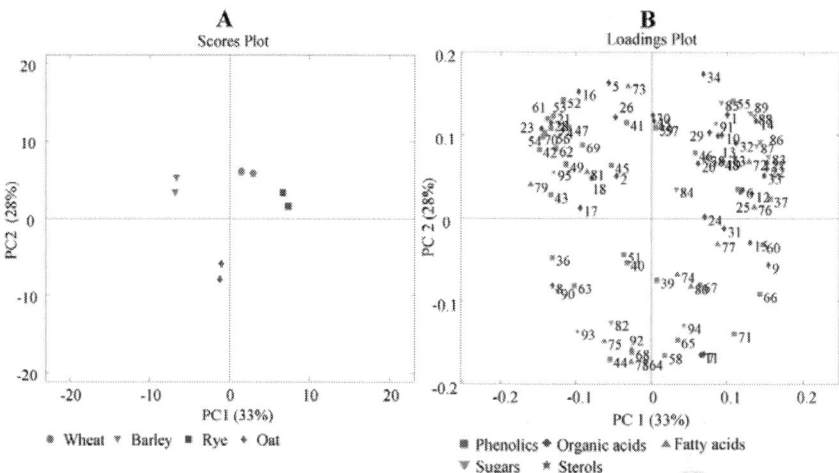

Figure 3. (**A**) scores and (**B**) loading plots of the three component PCA model developed using identified metabolite table. * Numbers in loadings plot correspond to the metabolites represented in Table 1.

3.3. Variation of Phenolics and Organic Acids in Cereals

Phenolic acid composition of wheat, barley, rye and oat were compared to previously reported data [18,21,22,23]. Figure 4 shows relative percentages of the nine most abundant, free and conjugated phenolic acids of cereals reported in previous studies and makes comparisons with the data obtained in the current study. In previous studies, the phenolic acids of cereals were extracted using 80% ethanol followed by hydrolysis of conjugated phenolics in 2 M sodium hydroxide solution and analyzed by LC-DAD. In the current study, free and conjugated phenolics were extracted using 85% methanol, hydrolyzed in 2 M solution of hydrochloric acid followed by GC-MS analysis and PARAFAC2 based data processing. These two methodologies in phenolic profiling of cereals result in several apparent compositional differences. However, it should be underlined that the compared cereal genotypes are different in the two studies and the goal of this study is not a comprehensive comparison of phenolics of cereal varieties, but to demonstrate the power of the standardized cereal metabolomics protocol developed.

Nine major phenolics of the cereals investigated in this study were compared with winter wheat (*Triticum aestivum* var. *aestivum*) [18], Dicktoo barley (USA) [21], Grandrieu rye (France) [23] and Bajka oat (Poland) [22] varieties (Figure 4). Figure 4 shows that the relative concentrations of caffeic acid consistently increased (14%–23%) in all cereal cultivars compared to the previous studies where its abundance was below 1%. Similarly, for wheat, barley and oat, concentrations of ferulic acid increased from approximately 20% to 33%, while the comparison is more consistent for the two rye varieties. These results suggest that in grains, a significant amount of caffeic and ferulic acids are present in conjugated forms that cannot be cleaved by alkaline hydrolysis. Thus, the most abundant phenolic acids in previous cereal metabolomics studies were ferulic, sinapinic and 3,5-dihydroxybenzoic acids, while in this study, ferulic, sinapinic and caffeic acids were the most abundant ones.

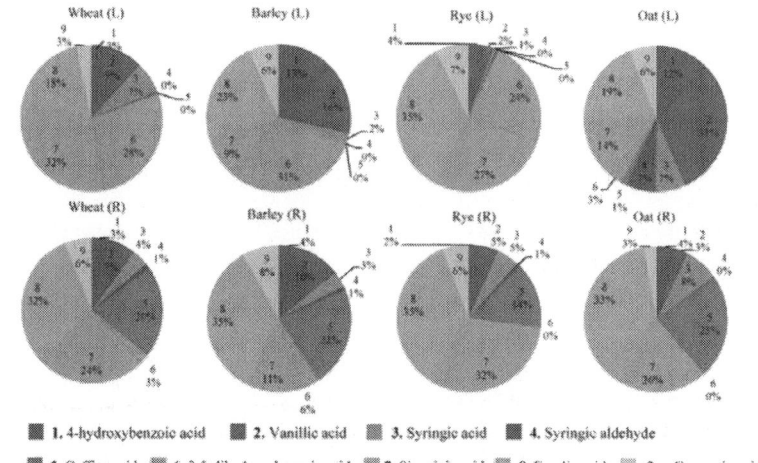

Wheat (L)
Barley (L)
Rye (L)
Oat (L)
Wheat (R)
Barley (R)
Rye (R)
Oat (R)
1. 4-hydroxybenzoic acid
2. Vanillic acid
3. Syringic acid
4. Syringic aldehyde
5. Caffiec acid
6. 3,5-dihydroxybenzoic acid
7. Sinapinic acid
8. Ferulic acid
9. p-Coumaric acid

Figure 4. Comparison of relative percentages of the nine most abundant phenolic acids of cereals reported in the literature (L) with the results of the current study (R). In literature the following genotypes were studied: winter wheat (*Triticum aestivum* var. *aestivum*) [18], Dicktoo barley (USA) [21], Grandrieu rye (France) [23] and Bajka oat (Poland) [22].

Wheat
Barley
Rye
Oat
18. Resorcinol
21. 3-hydroxyanthranilic ac. met.est.
24. Maseptol
31. 4-hydroxybenzeneacetic ac. met.est.
32. Vanillin
35. 4-hydroxyphenyl ethanol.
36. Anozol
40. 4-hydroxybenzoic acid
41. Methyl isovanillate
43. Syringaldehyde
46. 3,5-dihydroxybenzoic ac. met.est.
47. 2,5-dimethoxymandelic ac.
48. Vanillic acid
49. 4-hydroxycinnamic ac. met.est.
52. Methyl 2-(oxy)-2-(4-(oxy)phenyl)propanoate
54. 3,5-dihydroxy benzoic ac.
55. 3,4-dihydroxy benzoic ac.
58. Catechin1
59. Homovanillic acid
61. Catechin2
62. 2,5-dihydroxy benzoic ac.
64. Syringic acid
69. p-Coumaric acid
70. Ferulic ac. met.est.
71. 3,4,5-trihydroxy benzoic ac.
72. 2-hydroxymandelic ac. met.est.
74. Caffeic ac. met.est.
77. Ferulic acid
79. Sinapinic ac. met.est.
80. Methyl vanillactate
81. Caffeic acid
85. Sinapinic acid
0 0.02 0.04 0.06
Peak Ints (a.u)

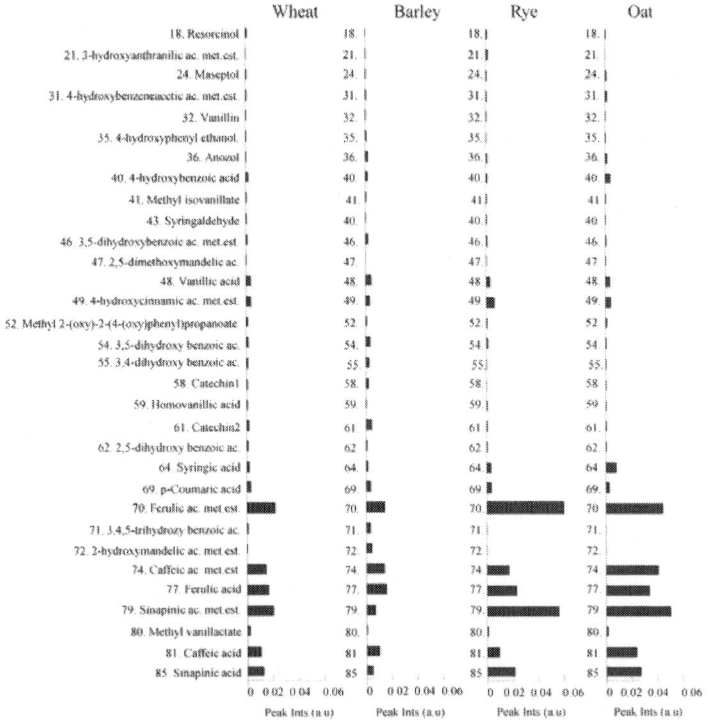

Figure 5. Relative concentrations of 32 phenolics detected from wheat, barley, rye and oat. Metabolites are numbered according to the Table 1.

3. RESULTS AND DISCUSSION

Figure 5 and Figure 6 demonstrate relative concentrations of phenolics and organic acids/alcohols of wheat, barley, rye and oat genotypes investigated in this study. Figure 5 show that ferulic, caffeic and sinapic acids and their methyl esters are the most abundant metabolites among all other phenolics in the cereal samples. Moreover, the relative concentrations of the most abundant phenolics are found to be up to three times greater in rye and oat than in wheat and barley. Succinic and 3-hydroxybutanoic acids were the most abundant metabolites among all organic acids detected in the four different cereals (Figure 5). Relative concentrations of fumaric and 2-hydroxycyclohexanecarboxylic acids were significantly higher in rye, while concentrations of malic and ketoglutaric acids were highest in barley.

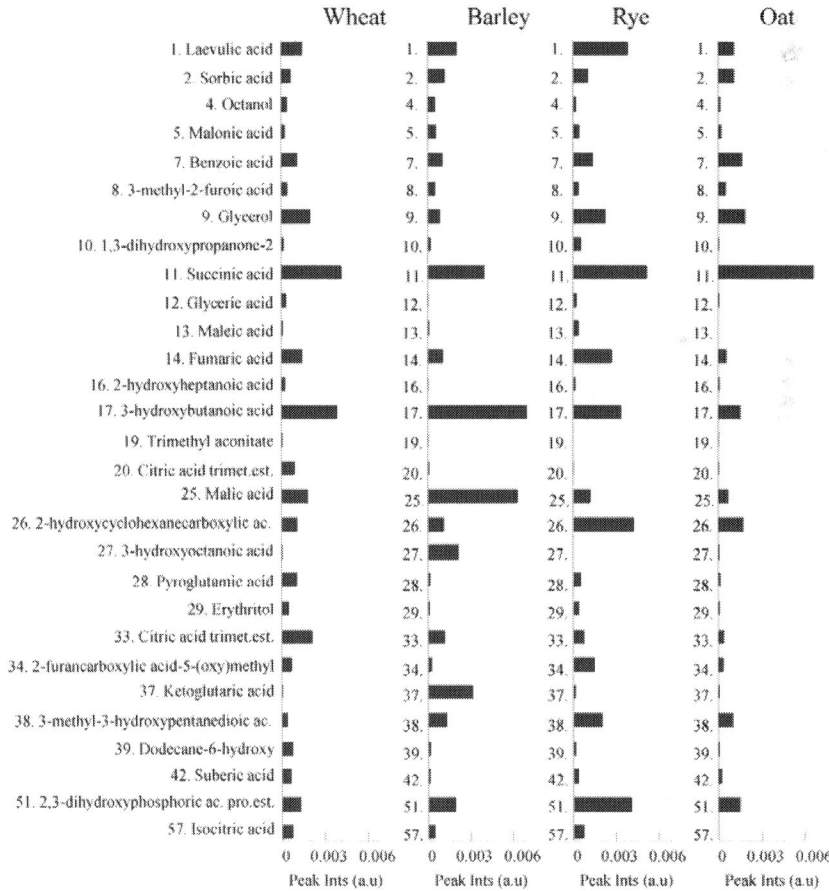

Figure 6. Relative concentrations of 29 organic acids/alcohols detected from wheat, barley, rye and oat. Metabolites are numbered according to the Table 1.

4. CONCLUSIONS

This paper outlines and demonstrates an optimized, relatively unbiased, comprehensive and high-throughput metabolomic profiling of whole-grain cereals based on new technologies developed within GC-MS metabolomics and chemometrics. A metabolite extraction protocol optimized towards phenolics and organic acids of whole-grains, and an unbiased and high-throughput protocol, was developed that allow processing of up to 60 samples per day. The hydrochloric acid based hydrolysis allowed extraction of all major cereal phenolics, free and conjugated, and enabled the detection of 32 phenolic and 30 organic acids from 50 mg of flour. A novel trimethylsilylation method based on TMSCN allowed the detection of up to 300 metabolites from the GC-MS profiles. The multi-way decomposition method PARAFAC2 facilitated deconvolution of overlapping, retention time shifted and low s/n ratio peaks with high precision and in a semi-automated manner. The resolved mass spectra of deconvoluted peaks allowed the identification of 89 metabolites using NIST and Golm metabolite databases. Multivariate and univariate analysis of phenolic profiles of cereals revealed that ferulic, caffeic and sinapinic acids and their esters were the main phenolics of whole-grain samples across the four cereals studied. Rye and oat showed higher concentrations of the most abundant phenolics acids compared to wheat and barley. Comparison of the relative concentrations of the nine most abundant phenolics of cereals with previously reported data showed that the acidic hydrolysis significantly improved detection of caffeic acid. However, metabolite profiles of cereals highly depend on several factors such as genotype, growth conditions, harvest time and storage. Thus, essential secondary metabolite profile comparisons of different cereals as well as different varieties require a strictly controlled experimental design. This paper has demonstrated a new methodology that is ready to be applied in a larger metabolomic profiling studies that may reveal biological information related to phenolic and organic acids of whole-grain cereals. Moreover, the protocol developed can easily be modified for polar metabolite fractions, including mono- and di-saccharides and amino acids, of cereals by altering metabolite extraction method and the additional of a methoximation step in GC-MS derivatization.

ACKNOWLEDGEMENTS

Faculty of Science is acknowledged for support to the elite-research area "metabolomics and bioactive compounds" with a PhD stipendium to B. Khakimov and The Ministry of Science and Technology is acknowledged for a grant to University of Copenhagen (S.B. Engelsen) with the title "metabolomics infrastructure" under which the GC-MS was acquired.

AUTHOR CONTRIBUTIONS

B.K. B.M.J. and S.B.E. designed the study; B.K. conducted the GC-MS analysis. B.K. and S.B.E. performed the chemometric analysis and drafted the manuscript. All authors contributed to, read and approved the final manuscript.

REFERENCES

1. Zilic, S.; Sukalovic, V.H.T.; Dodig, D.; Maksimovic, V.; Maksimovic, M.; Basic, Z. Antioxidant activity of small grain cereals caused by phenolics and lipid soluble antioxidants. *J. Cereal Sci* **2011**, *54*, 417–424.
2. Björck, I.; Östman, E.; Kristensen, M.; Anson, N.M.; Price, R.K.; Haenen, G.R.M.M.; Havenaar, R.; Knudsen, K.E.B.; Frid, A.; Mykkänen, H.; *et al.* Cereal grains for nutrition and health benefits: Overview of results from *in vitro*, animal and human studies in the HEALTHGRAIN project. *Trends Food Sci. Technol.* **2012**, *25*, 87–100.
3. Andersson, A.A.M.; Andersson, R.; Piironen, V.; Lampi, A.M.; Nystrom, L.; Boros, D.; Fras, A.; Gebruers, K.; Courtin, C.M.; Delcour, J.A.; *et al.* Contents of dietary fibre components and their relation to associated bioactive components in whole grain wheat samples from the HEALTHGRAIN diversity screen. *Food Chem.* **2013**, *136*, 1243–1248.
4. Amarowicz, R.; Zegarska, Z.; Pegg, R.B.; Karamac, M.; Kosinska, A. Antioxidant and radical scavenging activities of a barley crude extract and its fractions. *Czech J. Food Sci.* **2007**, *25*, 73–80.
5. Wood, P.J. Cereal beta-glucans in diet and health. *J. Cereal Sci.* **2007**, *46*, 230–238.
6. Mcintosh, G.H.; Whyte, J.; Mcarthur, R.; Nestel, P.J. Barley and wheat foods—Influence on plasma-cholesterol concentrations in hypercholesterolemic men. *Am. J. Clin. Nutr.* **1991**, *53*, 1205–1209.
7. Madhujith, T.; Shahidi, F. Antioxidative and antiproliferative properties of selected barley (*Hordeum vulgarae* L.) cultivars and their potential for inhibition of low-density lipoprotein (LDL) cholesterol oxidation. *J. Agric. Food Chem.* **2007**, *55*, 5018–5024.
8. Behall, K.M.; Scholfield, D.J.; Hallfrisch, J. Diets containing barley significantly reduce lipids in mildly hypercholesterolemic men and women. *Am. J. Clin. Nutr.* **2004**, *80*, 1185–1193.

9. Gibson, S.M.; Strauss, G. Implication of phenolic-acids as texturizing agents during cooking-extrusion cereals. *Abstr. Pap. Am. Chem. Soc.* **1991**, *202*, 150.
10. Vinson, J.A.; Erk, K.M.; Wang, S.Y.; Marchegiani, J.Z.; Rose, M.F. Total polyphenol antioxidants in whole grain cereals and snacks: Surprising sources of antioxidants in the US diet. *Abstr. Pap. Am. Chem. Soc.* **2009**, *238*, 246.
11. Khakimov, B.; Bak, S.; Engelsen, S.B. High-throughput cereal metabolomics: Current analytical technologies, challenges and perspectives. *J. Cereal Sci.* **2014**, *59*, 393–418.
12. Soltesz, A.; Smedley, M.; Vashegyi, I.; Galiba, G.; Harwood, W.; Vagujfalvi, A. Transgenic barley lines prove the involvement of TaCBF14 and TaCBF15 in the cold acclimation process and in frost tolerance. *J. Exp. Bot.* **2013**, *64*, 1849–1862.
13. Widodo; Patterson, J.H.; Newbigin, E.; Tester, M.; Bacic, A.; Roessner, U. Metabolic responses to salt stress of barley (*Hordeum vulgare* L.) cultivars, Sahara and Clipper, which differ in salinity tolerance. *J. Exp. Bot.* **2009**, *60*, 4089–4103.
14. Manavalan, L.P.; Chen, X.; Clarke, J.; Salmeron, J.; Nguyen, H.T. RNAi-mediated disruption of squalene synthase improves drought tolerance and yield in rice. *J. Exp. Bot.* **2012**, *63*, 163–175.
15. Balmer, D.; Flors, V.; Glauser, G.; Mauch-Mani, B. Metabolomics of cereals under biotic stress: Current knowledge and techniques. *Front. Plant Sci.* **2013**, *4*, 82.
16. Fernie, A.R.; Schauer, N. Metabolomics-assisted breeding: A viable option for crop improvement? *Trends Genet.* **2009**, *25*, 39–48.
17. Bino, R.J.; Hall, R.D.; Fiehn, O.; Kopka, J.; Saito, K.; Draper, J.; Nikolau, B.J.; Mendes, P.; Roessner-Tunali, U.; Beale, M.H.; *et al.* Potential of metabolomics as a functional genomics tool. *Trends Plant Sci.* **2004**, *9*, 418–425.
18. Li, L.; Shewry, P.R.; Ward, J.L. Phenolic acids in wheat varieties in the HEALTHGRAIN diversity screen. *J. Agric. Food Chem.* **2008**, *56*, 9732–9739.
19. Fernandez-Orozco, R.; Li, L.; Harflett, C.; Shewry, P.R.; Ward, J.L. Effects of environment and genotype on phenolic acids in wheat in the HEALTHGRAIN diversity screen. *J. Agric. Food Chem.* **2010**, *58*, 9341–9352.

20. Shewry, P.R.; Piironen, V.; Lampi, A.M.; Edelmann, M.; Kariluoto, S.; Nurmi, T.; Fernandez-Orozco, R.; Ravel, C.; Charmet, G.; Andersson, A.A.M.; *et al.* The HEALTHGRAIN wheat diversity screen: Effects of genotype and environment on phytochemicals and dietary fiber components. *J. Agric. Food Chem.* **2010**, *58*, 9291–9298.
21. Andersson, A.A.M.; Lampi, A.M.; Nystrom, L.; Piironen, V.; Li, L.; Ward, J.L.; Gebruers, K.; Courtin, C.M.; Delcour, J.A.; Boros, D.; *et al.* Phytochemical and dietary fiber components in barley varieties in the HEALTHGRAIN diversity screen. *J. Agric. Food Chem.* **2008**, *56*, 9767–9776.
22. Shewry, P.R.; Piironen, V.; Lampi, A.M.; Nystrom, L.; Li, L.; Rakszegi, M.; Fras, A.; Boros, D.; Gebruers, K.; Courtin, C.M.; *et al.* Phytochemical and fiber components in oat varieties in the HEALTHGRAIN diversity screen. *J. Agric. Food Chem.* **2008**, *56*, 9777–9784.
23. Nyström, L.; Lampi, A.M.; Andersson, A.A.M.; Kamal-Eldin, A.; Gebruers, K.; Courtin, C.M.; Delcour, J.A.; Li, L.; Ward, J.L.; Fras, A.; *et al.* Phytochemicals and dietary fiber components in rye varieties in the HEALTHGRAIN diversity screen. *J. Agric. Food Chem.* **2008**, *56*, 9758–9766.
24. Ward, J.L.; Poutanen, K.; Gebruers, K.; Piironen, V.; Lampi, A.M.; Nystrom, L.; Andersson, A.A.M.; Aman, P.; Boros, D.; Rakszegi, M.; *et al.* The HEALTHGRAIN cereal diversity screen: Concept, results, and prospects. *J. Agric. Food Chem.* **2008**, *56*, 9699–9709.
25. Arranz, S.; Calixto, F.S. Analysis of polyphenols in cereals may be improved performing acidic hydrolysis: A study in wheat flour and wheat bran and cereals of the diet. *J. Cereal Sci.* **2010**, *51*, 313–318.
26. Sani, I.M.; Iqbal, S.; Chan, K.W.; Ismail, M. Effect of acid and base catalyzed hydrolysis on the yield of phenolics and antioxidant activity of extracts from germinated brown rice (GBR). *Molecules* **2012**, *17*, 7584–7594.
27. Khakimov, B.; Motawia, M.S.; Bak, S.; Engelsen, S.B. The use of trimethylsilyl cyanide derivatization for robust and broad-spectrum high-throughput gas chromatography-mass spectrometry based metabolomics. *Anal. Bioanal. Chem.* **2013**, *405*, 9193–9205.
28. Bro, R.; Andersson, C.A.; Kiers, H.A.L. PARAFAC2—Part II. Modeling chromatographic data with retention time shifts. *J. Chemom.* **1999**, *13*, 295–309.
29. Amigo, J.M.; Skov, T.; Coello, J.; Maspoch, S.; Bro, R. Solving GC-MS problems with PARAFAC2. *Trac-Trends Anal. Chem.* **2008**, *27*, 714–725.

30. Khakimov, B.; Amigo, J.M.; Bak, S.; Engelsen, S.B. Plant metabolomics: Resolution and quantification of elusive peaks in liquid chromatography-mass spectrometry profiles of complex plant extracts using multi-way decomposition methods. *J. Chromatogr. A* **2012**, *1266*, 84–94.
31. Vandendool, H.; Kratz, P.D. A generalization of retention index system including linear temperature programmed gas-liquid partition chromatography. *J. Chromatogr.* **1963**, *11*, 463.
32. Golm Metabolome Database. Available online: http://gmd.mpimp-golm.mpg.de/ (accessed on 5 November 2013).
33. Hotelling, H. Analysis of a complex of statistical variables into principal components. *J. Educ. Psychol.* **1933**, *24*, 417–441.
34. Sumner, L.; Amberg, A.; Barrett, D.; Beale, M.; Beger, R.; Daykin, C.; Fan, T.; Fiehn, O.; Goodacre, R.; Griffin, J.; *et al*. Proposed minimum reporting standards for chemical analysis.*Metabolomics* **2007**, *3*, 211–221.

CHAPTER 6

Discrimination of Wild-Grown and Cultivated Ganoderma lucidum by Fourier Transform Infrared Spectroscopy and Chemometric Methods

Ying Zhu[1]*, Augustine Tuck Lee Tan[2]

[1]Mathematics and Mathematics Education, National Institute of Education, Nanyang Technological University, Singapore
[2]Natural Sciences and Science Education, National Institute of Education, Nanyang Technological University, Singapore

ABSTRACT

Wild-grown Ganoderma lucidum (G. lucidum), a traditional Chinese herbal medicine, is highly cherished and expensive for its medicinal efficiency. This study targets the development of an accurate and effective analytical method to distinguish wild-grown G. lucidum from cultivated ones, which are of essential importance for the quality assurance and estimation of its medicinal value. Furthermore, different parts of G. lucidum have been studied to examine the differences between wild-grown and cultivated ones. Fourier transform infrared (FTIR) diffuse reflectance spectroscopy combined with the appropriate chemometric method has been proven to be a rapid and powerful tool for discrimination of wild-grown and cultivated G. lucidum with classification accuracy of 98%. The informative spectral absorption bands for discrimination emphasized by the linear diagnostic rule have provided quantitative interpretations of the chemical constituents of wild- grown G. lucidum regarding its anticancer effects.

Keywords: Ganoderma lucidum, Traditional Chinese Medicine, Fourier Transform Infrared Spectroscopy, Chemometrics, Principal Component Discriminant Analysis, Partial Least Squares Discriminant Analysis

1. INTRODUCTION

Ganoderma lucidum (G. lucidum), a traditional Chinese herbal medicine called "lingzhi" in Chinese, has been widely used as a medical remedy in China and other East Asian countries for centuries. According to the literature record of Chinese medical classic, Herbal Compendium of Shen Nong, this edible mushroom is one of the most esteemed and potent herbal medicines used for maintaining good health and preventing disorders and diseases. It has also been considered as a potential candidate for treatment of different diseases, including cancer [1] [2].

The wild-grown high-quality G. lucidum is rare in nature, and thus has always been highly cherished and expensive for its medicinal value. The cultivated G. lucidum has been commercially in demand, particularly in China during the past several decades. Due to their different growing conditions, the wild-grown and cultivated G. lucidums may contain different levels of effective chemical components which affect their quality and medicinal efficacy. Since many G. lucidum products now come in various formulations such as capsules and powder, it is difficult to identify its wild-grown product by means of physical appearance, smell, or taste. Therefore, an accurate and effective analytical method to determine the differences between wild and cultivated G. lucidum in their unprocessed states is of essential importance for the quality assurance and estimation of medicinal value before it is converted to the final product.

G. lucidum, like other Chinese herbal medicines, is a complicated system of compounds. Currently, herbal medicines have been commonly investigated with the use of high performance liquid chromatography (HPLC), thin layer chromatography (TLC), and colorimeter. These methods are found to be expensive, time-consuming, labour-intensive, and requiring a large quantity of organic solvents. Also, the results are inadequate for classification purposes because of the limited amount of active chemical components that can be detected in what is a very complex system [3] [4].

Fourier transform infrared (FTIR) spectroscopic methods have many advantages for the classification of herbal medicines in terms of easy and direct usage of technique, non-destructiveness, a small quantity of samples needed and short data acquisition time. Studies on herbal medicines using the FTIR technique are still in its infancy [5] [6].

Furthermore, FTIR spectra of herbal medicines consist of many overlapping absorption bands representing the different modes of vibration of a large number of molecular constituents in the compounds. These vibrational bands are sensitive to the physical and chemical states of the compounds, and they can be detected at low levels [4]. However, the differences in the FTIR spectra within the same herbal species may be subtle and even not visible to the naked eye. Even for experienced analysts, distinguished by simple visual inspection, the slight difference between samples among particular absorption bands is subjective and the results may vary between analysts. Therefore, suitable chemometric methods have been applied in our study to analyze the FTIR spectra.

For the analysis of herbal medicine, there are limited studies which quantify the main constituents in herbal medicines samples, like ginseng, semen cassia, and G. lucidum, by using FTIR spectroscopy [3] - [6] . Moreover, to our best knowledge, there have been no such reports on the discrimination of wild-grown G. lucidum samples from cultivated ones using FTIR spectroscopy.

This paper investigated the feasibility of the discrimination between wild-grown and cultivated G. lucidum, as well as the discrimination between different parts of the G. lucidum, by FTIR diffuse spectroscopy along with chemometric methods. The multivariate methods based on linear discriminant analysis (LDA) explored in this paper would be simple, robust and computationally efficient. In particular, the directions of linear discriminant vectors can be potentially interpretable as directions with the informative spectral bands emphasized for discrimination, which would be useful in exploring the correlations between spectral features and the major chemical compounds of wild-grown G. lucidum regarding its anti-cancer effects.

2. MATERIALS AND METHODS

2.1. Sample Preparation

The cultivated G. lucidum and the wild-grown G. lucidum were originated from Taishan, China. The fruiting body (or pileus) of both types of G. lucidum was cross-sectioned into thin slices. A cross-section of the G. lucidum slice showed three structured layers with colours growing lighter from the top to the bottom of the pileus: the upper crust (skin), the mid-context layer (flesh), and the lower tubular layer (fine channels) as shown in Figure 1. A total of 15 cultivated sliced samples and 15 wild-grown sliced samples were used for data collection. For each sample, four parts (top surface, upper middle area, lower middle area and bottom surface) were studied. From each part, three spectra were collected at different positions with one from centre, one from right-hand side and one from left-hand side of the cross-section, as illustrated in Figure 1. In total, 360 G. lucidum spectra, including 180 spectra from cultivated samples and 180 spectra from wild-grown samples, were then collected and used for our analysis.

Before collecting the FTIR diffuse reflectance spectrum, the fine powder of each raw sample was transferred into a circular (1-cm diameter) silicon carbide (SiC) disc by rubbing it on the sample. The diffuse reflectance spectrum of the G. lucidum powder coating on the disc was recorded directly without further processing of the sample.

2.2. FTIR Spectroscopic Measurement

A FTIR spectrometer (Perkin Elmer Model 100) equipped with a diffuse reflectance accessory was used to record the diffuse reflectance spectra of the G. lucidum powder coated on SiC discs. The FTIR diffuse reflectance spectra were recorded in the mid-IR region of 4000 - 400 cm^{-1} at resolution of 4 cm^{-1} with 16 scans for each spectrum. Each spectrum with high signal-to-noise signal was obtained by an average of these 16 scans. The background spectrum which was the diffuse reflectance spectrum of the SiC disc without the sample powder was also recorded with the same parameters. The sample spectrum was then ratioed with the background spectrum to obtain a transmittance or absorbance spectrum with the unwanted absorption bands of water and carbon dioxide removed [7]. Therefore, the diffuse reflectance absorption spectrum of a G. lucidum sample with strong absorption bands was accurately collected.

2.3. Spectral Pre-Treatment

FTIR spectra are affected by both the concentration of the chemical constituents and the physical properties of the analyzed product, and the latter properties account for the majority of the variance among spectra while the variance due to chemical composition is considered to be small. It is necessary to perform mathematical pre- treatments to reduce the variation due to physical effects, such as baseline variation, light scattering, path length differences, etc, so as to enhance the contribution of the chemical composition [4] [6] [8].

The spectra were first smoothed using the Savitzky-Golay algorithm [9], spanning a 10-point window. To speed subsequent manipulation, the smoothed data were then reduced by taking every third point only. To remove the regions of the spectra with low signal-to-noise ratios arising from the lower system response, only the wavenumbers ranging from 4000 to 450 cm^{-1}, with 593 spectra points at 5.987 cm^{-1} intervals, were used in the analysis. The standard normal variate (SNV) method [10], as a mathematical transformation method for spectra, was used to remove slope variation and to correct light scatter due to different particle sizes. The spectra were therefore normalized by setting the mean intensity of each spectrum to zero and the variance to one. The mean spectra from cultivated and wild-grown G. lucidum after pretreatment were presented in Figure 2.

2.4. Statistical Analysis

FTIR spectra of herbal medicines consist of many overlapping absorption bands which are the product of complex patterns of biochemical components. Multivariate statistical methods including principal component analysis (PCA), partial least squares (PLS) and linear discriminant analysis (LDA) [11] - [13] were therefore employed in this study to investigate the differences of spectra from wild-grown and cultivated G. lucidum. PCA and PLS were used to reduce the dimension of the original spectral data matrix, X, with little loss of information. From a large number of variables measured on a given set of samples, PCA extracts a small to moderate number of new variables that account for most of the variability between samples. The new variables, called principal components (PCs), are linear combinations of all the original spectral measurements and are uncorrelated to each other. Alternatively, PLS seeks to find a small to modest number of latent variables, each of which, called PLS component, is obtained by maximizing the covariance between response y and all possible linear functions of X. Then LDA focuses on finding a linear combination of the new variables, provided either by PCA or PLS, to construct canonical variate which best separates the two groups. Using pretreated spectra data described in Section 2.3, classification rules were derived using principal component discriminant analysis (PCDA) [14] [15], and partial least squares discriminant analysis (PLSDA) [15] [16]. The PCDA involved an initial PCA on the pre-treated spectra followed by a LDA performed on the first k PCs' scores. The PLSDA involved a PLS regression on the pre-treated spectra followed by a LDA on the first k PLS components' scores. Both PCDA and PLSDA were carried out with k ranging from 2 to 20.

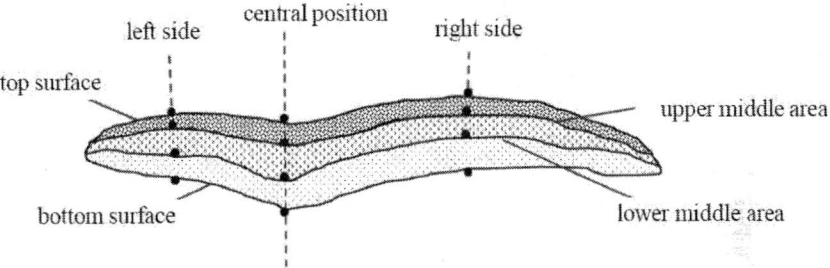

Figure 1. A cross-section of fruiting body of G. lucidum.

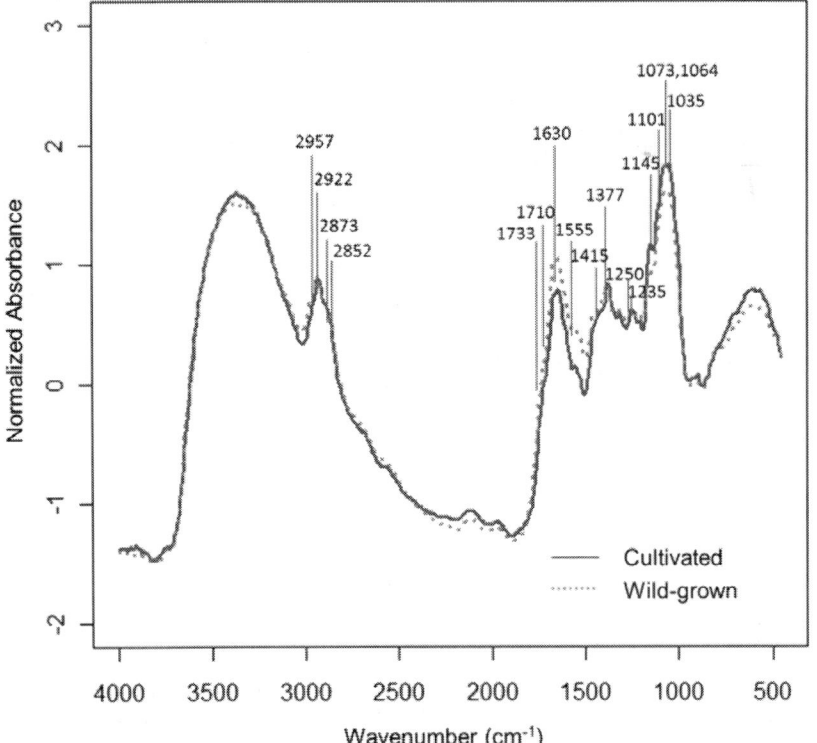

Figure 2. Mean spectra from cultivated (solid line) and wild-grown (dotted line) G. lucidum after standard pre-treatment.

Leave-one-out cross-validation was used to train the algorithm by carrying out the PCDA or PLSDA classification rule on all the data except one site which was then tested. This was repeated until all sites have been tested and an overall model accuracy was determined. To ensure that the results obtained are not training set specific, a repeated holdout validation as an alternative analysis was used with 60% of the data to train and 40% to test the model. To ensure statistical robustness, this process was repeated 50 times with different randomly resampled training and test sets, and the averages with their standard deviation were presented to assess the classification performance.

Furthermore, since different parts of G. lucidum showed very different internal structures, the samples collected from different parts of the fruiting body were believed to have different constituent properties [17] . The same PCDA, PLSDA procedure and validation analysis were carried out on wild-grown and cultivated G. lucidum for discrimination between spectra from different parts of the pileus.

Since a preliminary study showed that the spectra collected from the different positions (central, left or right) within the same part of the sample do not show significant differences for either cultivated or wild-grown G. lucidum, which can also been supported by the results in Section 3.3, hereby we treated those spectra from the same part of the sample as the ones from the same site. Therefore, all the above analysis was carried out on a per-site-base since this usually gave more reliable results than a per-spectra analysis. In per-site analysis, the linear discrimination rule was based on the average canonical scores of the spectra from one particular site for either leave-one-out cross-validation or repeated holdout validation.

All the algorithms for computations and analyses were implemented in R statistical programming language [18] .

3. RESULTS AND DISCUSSION

3.1. Absorption Band Assignments of FTIR Spectra of G. lucidum

The typical FTIR spectrum of G. lucidium after pretreatment was presented in the region of 4000 - 450 cm^{-1}. The major peaks of the absorption bands were labelled on the mean spectrum as shown in Figure 2. Table 1 provided the wavenumbers and their corresponding assignments of the aborption bands in the FTIR spectrum of G. lucidum based on literature [19] - [22].

The polysaccharides and the triterpene compounds (also known as triterpenoids) have long been established to be the most biologically active substances in G. lucidum [23]. The bioactive polysaccharides in the forms of glucomannan and arabinan identified by the absorption band at 1064 cm^{-1} and 1035 cm^{-1} respectively (listed in Table 1) in G. lucidum have been demonstrated to exhibit strong anti-tumor activities including preventing oncogenesis and tumor metastasis [24] [25] . Furthermore, the triterpene compounds identified by the absorption band at 1377 cm^{-1} and 1145 cm^{-1} (given in Table 1) in G.

lucidum, has shown to inhibit primary solid-tumor growth in the spleen and secondary metastatic tumor growth in the liver [24].

3.2. Discrimination between Wild-Grown and Cultivated G. lucidum

As shown in Figure 2, the mean spectra of wild-grown and cultivated G. lucidum after pretreatment had very similar patterns. It was difficult to distinguish between wild-grown and cultivated G. lucidum through visual inspection, which indicated that the major components in two types of G. lucidum are similar. Thus, appropriate multivariate chemometric methods as described in Section 2.4 were applied to discern the differences between these two types of G. lucidum.

For PCDA model, the number of PCs chosen is crucial to the performance of discrimination. The discrimination results of cross-validation and repeated holdout were used to optimize the number of PCs. The first fifteen PCs representing 98% of the total variance in the spectral data were used to construct the PCDA model for discriminating between wild-grown and cultivated G. lucidum. Results of the leave-one-out cross-validation and the repeated holdout validation analyses for assessing the PCDA model were shown in Table 2(a). The leave- one-out cross-validation analysis gave a discrimination accuracy of 98%. The repeated holdout validation analysis produced an average discrimination accuracy of 96% with 2% standard deviation.

With the relationship between the spectra variables and the responses taken into account for latent variable design, the PLSDA model appeared to do a better job than the PCDA model as shown in Table 3(a), giving comparable discrimination results but using a fewer optimal number of latent variables (only four) in constructing the canonical variate.

The results from both models suggested that there may exist some inherent compositional differences caused by different growing environment between cultivated and wild-grown G. lucidum even though they actually belong to the same species.

3.3. Discrimination between Different Parts of G. lucidum Slice

When studying on the differences between different parts of G. lucidum slice, we combined the spectral data of upper middle area and that of lower middle area. So three different parts, top, middle and bottom parts, of pileus of G. lucidum, were studied to examine their differences.

The PCDA model was firstly applied to discriminate between top and middle parts of pileus for cultivated and wild-grown G. lucidum separately. The two groups, cultivated and wild-grown G. lucidum, showed comparably good discrimination results with above 98% accuracy for leave-one-out cross-validation and 98% accuracy for repeated holdout validation analysis as shown in Table 2(b) and Table 2(c).

Table 1. Absorption band assignments of the FTIR spectrum of G. lucidum.

Wavenumber (cm^{-1})	Functional Group Assignments
2957	CH$_3$asym stretching (mainly lipids)
2922	CH$_2$asym stretching (mainly lipids)
2873	CH$_3$sym stretching (mainly proteins)
2852	CH$_2$sym stretching (mainly lipids)
1710, 1733	C=O carbonyl stretching of saturated aliphatic esters
1630	Amide I (protein C=O stretch)
1555	Amide II (C-N, N-H stretching) mainly proteins
1415	O-H bending, polysaccarides
1377	Symmetric bending of aliphatic CH$_3$, triterpene compounds (CH$_2$=CH-CH$_3$)
1250	Pectic substances
1235	Amide III (C-N, N-H stretching) mainly proteins
1145	Cellulose (β-glucan), triterpene compounds (C-O)
1101	Antisym in-phase, pectic substances
1073	Rhamnogalactorunan, β-galactan
1064	C-O stretching, polysaccharides (glucomannan)
1035	OH and C-OH stretching in sugars, polysaccharides (arabinan)

Table 2. Discrimination results of PCDA model using leave-one-out cross-validation and repeated holdout validation for differentiating between (a) spectra of cultivated and wild-grown G. lucidum; (b) spectra of different parts of wild-grown G. lucidum; (c) spectra of different parts of cultivated G. lucidum using optimal number of PCs for LDA.

Accuracy		Optimal Number of PCs	Leave-One-Out Cross-Validation	Repeated Holdout	(Standard Deviation)
(a) Wild vs Cultivated		15	98%	96%	(2%)
(b) Wild	Top vs Mid	5	98%	98%	(1%)
	Bottom vs Mid	7	99%	96%	(2%)
(c) Cultivated	Top vs Mid	5	100%	98%	(1%)
	Bottom vs Mid	7	87%	83%	(4%)

Table 3. Discrimination results of PLSDA model using leave-one-out cross-validation and repeated holdout validation for differentiating between (a) spectra of cultivated and wild-grown G. lucidum; (b) spectra of different parts of wild-grown G. lucidum; (c) spectra of different parts of cultivated G. lucidum using optimal number of PLS components for LDA.

Accuracy		Optimal Number of PLS Components	Leave-One-Out Cross-Validation	Repeated Holdout	(Standard Deviation)
(a) Wild vs Cultivated		4	98%	97%	(2%)
(b) Wild	Top vs Mid	2	98%	98%	(1%)
	Bottom vs Mid	3	100%	97%	(2%)
(c) Cultivated	Top vs Mid	2	100%	98%	(1%)
	Bottom vs Mid	3	87%	84%	(3%)

3. Results and Discussion

When the PCDA model was then applied to discriminate between bottom and middle parts of pileus for cultvated and wild-grown G. lucidum separately, the two groups showed different discrimination performances. The wild-grown group achieved a high accuracy of 99% for leave-one-out cross-validation and 96% for repeated holdout validation while the cultivated group gave a lower accuracy of 87% for leave-one-out cross-validation and 83% for repeated holdout validation as presented inTable 2(b) and Table 2(c).

The PLSDA model gave fairly consistent results with PCDA model, but used a very small number of optimal latent variables (only two or three) for discrimination as shown in Table 3(b) and Table 3(c).

The results suggested that the differences between upper and middle parts were prominent for both wild- grown and cultivated G. lucidum. However, the differences between middle and bottom parts of wild-grown G. lucidum may be better detected than that of cultivated one. These findings can also be presented in a 3D space diagram provided by the PCDA or the PLSDA model. Figure 3(a) and Figure 3(c) showed that in the 3D space represented by the first three PCs, a clear separation plane can be found to discriminate top part from middle part for either wild-grown or cultivated group. When distinguishing between middle and bottom parts, there was a neat separation between the two parts for the samples from wild-grown group as illustrated in Figure 3(d). However, for the samples from cultivated group, the separation between the middle and bottom parts was not clearly displayed as shown in Figure 3(b). The discrimination between different parts of G. lucidum can be much better displayed in a 3D space by using PLSDA model as shown in Figure 4. With the first four PLS components used in constructing the optimal PLSDA model, the 3D scatter plot of the first three PLS components in Figure 4(d) illustrated a much clearer separation between middle and bottom parts of wild-grown G. lucidum, when compared to the corresponding 3D scatter plot of the first three PCs in Figure 3(d) which explained 90% of the total variance of the spectral data.

The mean spectra from middle and bottom parts of wild-grown G. lucidum also showed some increased between-group differences compared with that of cultivated one, especially at certain spectral regions like ~2900 cm^{-1}, ~1600 cm^{-1}, and ~1000 cm^{-1} as shown in Figure 5.

Further interest in the differences between central and side (left-side or right-side) positions within the same part of pileus were also explored by using a PCDA model with number of PCs varying from 7 to 15. No obvious difference was found for either cultivated or wild-grown G. lucidum as shown inTable 4. This also verified that it was reasonable to treat those spectra from the different positions of the same part of pileus as replicated spectra collected from the same site, as mentioned in Section 2.4.

With different growing environments, different parts of G. lucidum may contain different levels of major chemical components and thus show some different internal structures. Completely exposed to nature or cultivated environments, the upper part of G. lucidum often changes quickly. However, the changes of bottom part with environments usually take longer time, and thus those wild-grown one, which grows slower and are harvested when being old, showed more differences between bottom and middle parts than the cultivated one. The internal structure of the fruiting body of G. lucidum seems to be very important for identifying wild-grown group from cultivated one.

3.4. Correlation between Spectral Absorption Bands and Chemical Components of G. lucidum and Its Medicine Effect

Discrimination performance may be explained by the correlation between spectral feature and chemical constituents of G. lucidum. The PCDA loading combining the loadings from the PCA and LDA gives the PCDA loading of the original variables in constructing a canonical variate. In the same way as for the PCDA loading, combining the loadings from the PLS and LDA gives the PLSDA loading of the original variables. Both the PCDA loading and the PLSDA loading show the contribution at each wavelength to the linear diagnostic rule and thus can be related easily to the spectral features, which permits interpretation of its spectral basis. In Figure 6, the most obvious feature was a large PCDA loading in the regions of 1150 - 1000 cm^{-1} and 1760 - 1600 cm^{-1} (peaking at around 1000 cm^{-1} and 1600 cm^{-1} respectively) corresponding to some slight differences between the mean spectra of wild-grown and cultivated G. lucidum. The prominent absorption peaks observed in the region of 1150 - 1000 cm^{-1} were very characteristic of triterpenoids and polysaccharides due to the C-O and C-C vibrations. The other prominent absorption peaks at around 1760 - 1600 cm^{-1} were consistent with a C=O stretching vibration in carbonyl compounds which may be characterized by the presence of high content of terpenoids and protein in the mixture of G. lucidum. The presence of a sharp peak at ~2900 cm^{-1} was due to C-H stretching vibration [26] -[28] .

A comparison between the PCDA loadings and the PLSDA loadings for discrimination between cultivated and wild-grown G. lucidum can be seen in Figure 7. The loading features observed from PLSDA model were very similar to those from the PCDA model. The major features of these loadings can also be explained by the assignments of the corresponding absorption bands in the FTIR spectrum listed in Table 1. The high consistency between the loadings of the linear diagnostic rules and the chemical features of the FTIR spectrum may provide a quantitative explanation of the major chemical constituents of wild-grown G. lucidum with respect to chemometrics.

3. Results and Discussion

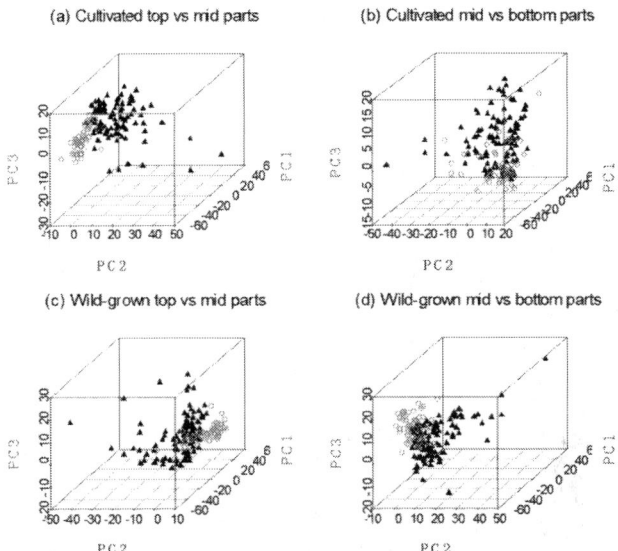

Figure 3. The 3D scatter plot of the first three PCs' scores of the spectra from (a) top and middle parts of cultivated G. lucidum; (b) middle and bottom parts of cultivated G. lucidum; (c) top and middle parts of wild-grown G. lucidum; (d) middle and bottom parts of wild-grown G. lucidum (top or bottom part: circle (○); middle part: triangle (▲)).

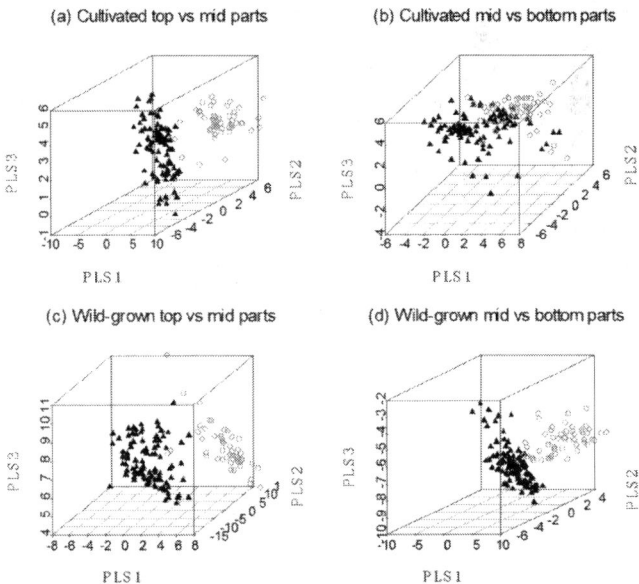

Figure 4. The 3D scatter plot of the first three PLS components' scores of the spectra from (a) top and middle parts of cultivated G. lucidum; (b) middle and bottom parts of cultivated G. lucidum; (c) top and middle parts of wild-grown G. lucidum; (d) middle and bottom parts of wild-grown G. lucidum (top or bottom part: circle (○); middle part: triangle (▲)).

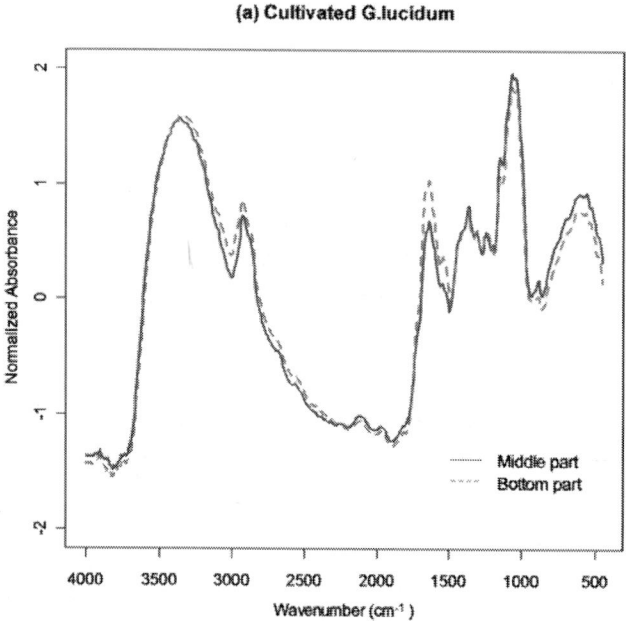

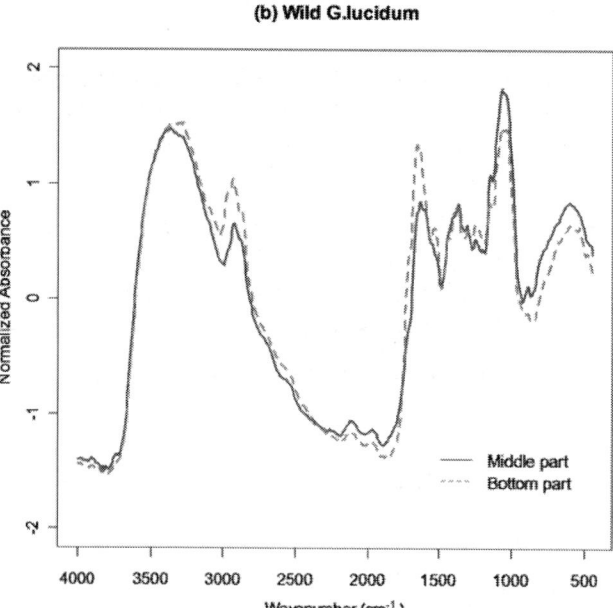

Figure 5. Mean spectra from middle part (in solid line) and bottom part (in dotted line) of (a) cultivated G. lucidum, and (b) wild-grown G. lucidum.

3. Results and Discussion

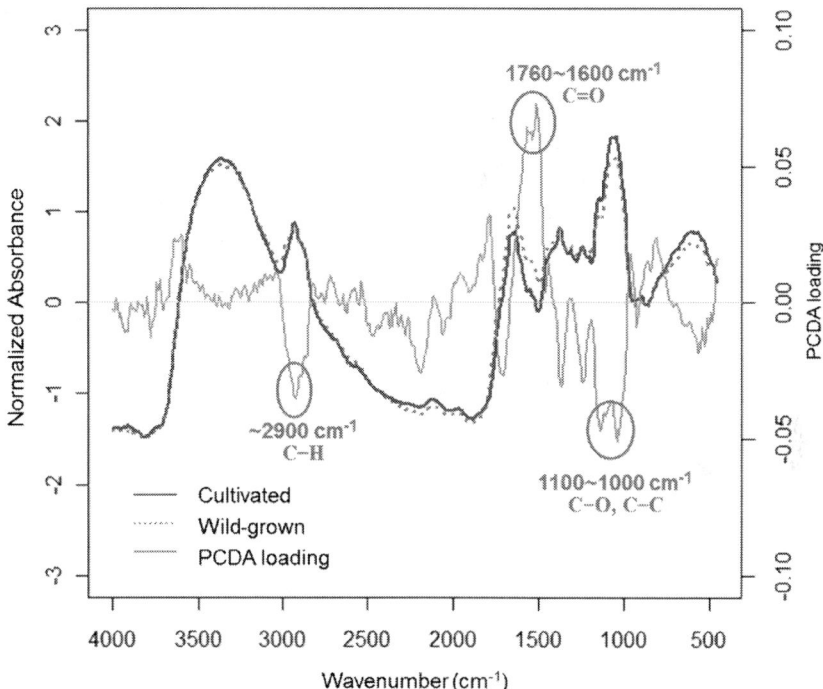

Figure 6. PCDA loadings for discrimination between the spectra from cultivated and wild-grown G. lucidum. The PCDA loading is shown in grey, with the mean spectra for the two types superimposed (cultivated in solid line, wild-grown in dotted line).

Table 4. Discrimination results of leave-one-out cross-validation for differentiating between central and side positions within the same part of pileus of (a) cultivated G. lucidum or (b) wild-grown G. lucidum, using seven to fifteen PCs for LDA.

Accuracy	Central Position vs Side Positions		
	Top Part	Middle Part	Bottom Part
(a) Cultivated	57.6% - 62.1%	51.5% - 56.1%	50.6% - 53.3%
(b) Wild-grown	52.5% - 55.6%	53.5% - 54.8%	53.6% - 58.5%

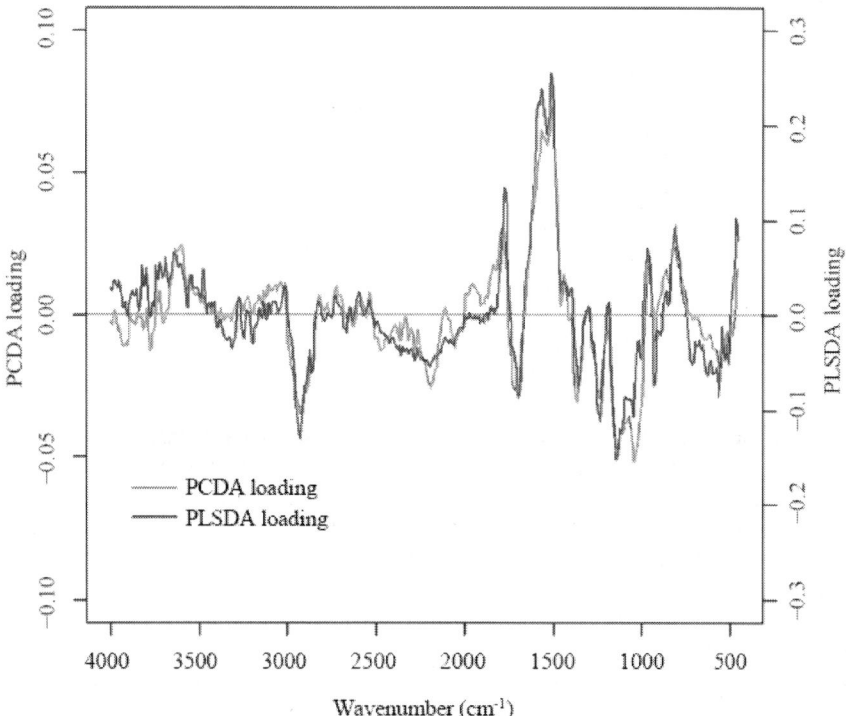

Figure 7. Comparison of PCDA loadings (green line) and PLSDA loadings (blue line) for discrimination between the spectra from cultivated and wild-grown G. lucidum.

It is known that G. lucidum contains approximately 400 different bioactive compounds [23]. Among these ingredients, triterpenoids, polysaccharides and protein are the major chemical constituents of G. lucidum [25]. These biologically active compounds have been demonstrated to prevent oncogenesis and tumor metastasis, and thus have anticancer effect [24] [25] [29] . Some comparative studies also reported that the different parts of the fruiting body of G. lucidum showed differences with regard to their antitumor effects in human breast cancer cells and immunomodulatory activities [27] .

4. CONCLUSIONS

Wild-grown G. lucidum is a rare and cherished herb for its many therapeutic effects and medicinal value. Since many G. lucidum products are sold in various formulations, distinguishing between wild-grown and cultivated G. lucidum products by morphological means becomes difficult. Thus the quality assurance of wild-grown G. lucidum is of essential importance.

In this study, FTIR spectroscopy combined with multivariate analysis after the appropriate spectral data pre- treatment has been proved to be a very powerful tool to distinguish wild-grown G. lucidum from cultivated ones. The great advantage of FTIR spectroscopy is its easiness of sample preparation, no need of sample destruction and rapid identification of natural products. The results of this study showed that an excellent classification performance can be obtained by linear discrimination models with accuracy up to 98%. Both the PCDA and the PLSDA model can achieve comparable classification accuracy. But the PLSDA model was simpler than the PCDA model by using a small number of latent variables, which had advantages in terms of algorithm implementation and model interpretation. Furthermore, different parts of G. lucidum have been studied to investigate the differences between wild-grown and cultivated ones. The wild-grown G. lucidum showed more differences than the cultivated one in its internal structure of the fruiting body, particularly between bottom and middle parts, which seemed to be very important for identifying wild-grown product.

These results suggested that this multivariate analysis method may have commercial and regulatory potential to avoid time-consuming recalibration work, costly and laborious chemical and visual analysis for each sample. Though using entire spectral band, the mathematical classification algorithm based on linear discriminant analysis was computationally efficient. Most importantly, the directions of discriminant vectors used here can be physically interpretable as directions where the informative spectral bands for classification are emphasized, which showed some correlation between spectral absorption band and certain important chemical components of wild-grown G. lucidum with anticancer effects.

This is a novel, yet interesting finding, as there have been no such studies on G. lucidum which have shown this correlation from the viewpoint of chemometrics. This work therefore played an important role of providing quantitative interpretation and scientific support to the claims on the health benefits of G. lucidum as well as on the antitumor properties of wild-grown G. lucidum. Further studies will focus on the different parts of G. lucidum for its medicinal value by using FTIR spectroscopy.

ACKNOWLEDGEMENTS

This work is funded by Academic Research Fund (AcRF: RI 12/10 TTL) of National Institute of Education, Nanyang Technological University, Singapore.

REFERENCES

1. Yuen, J.W. and Gohel, M.D. (2005) Anticancer Effects of Ganodermalucidum: A Review of Scientific Evidence. Nutrition and Cancer, 53, 11-17.
2. Lin, Z.B. and Zhang, H.N. (2004) Anti-Tumor and Immunoregulatory Activities of Ganoderma lucidum and Its Possible Mechanisms. Acta Pharmacologica Sinica, 25, 1387-1395.

3. Chew, O.S., Hamdan, M.R., Ismail, Z. and Ahmad, M.N. (2004) Assessment of Herbal Medicines by Chemometrics- Assisted Interpretation of FTIR Spectra. Analytica Chimica Acta, 570, 116-123.
4. Lai, Y.H., Ni, Y.N. and Kokot, S. (2010) Classification of Raw and Roasted Semen Cassiae Samples with the Use of Fourier Transform Infrared Fingerprints and Least Squares Support Vector Machines. Applied Spectroscopy, 64, 649- 656.
5. Chen, Y., Xie, M.Y., Yan, Y., Zhu, S.B., Nie, S.P., Li, C., Wang, Y.X. and Gong, X.F. (2008) Discrimination of Ganoderma lucidium According to Geographical Origin with Near Infrared Diffuse Reflectance Spectroscopy and Pattern Recognition Techniques. Analytica Chimica Acta, 618, 121-130.
6. Yap, K.Y.L., Chan, S.Y. and Lim, C.S. (2007) Authentication of Traditional Chinese Medicine Using Infrared Spectroscopy: Distinguishing between Ginseng and Its Morphological Fakes. Journal of Biomedical Science, 14, 265-273.
7. Smith, B.C. (1996) Fundamentals of Fourier Transform Infrared Spectroscopy. CRC Press, London.
8. Woo, Y.A., Kim, H.J. and Cho, J. (1999) Identification of Herbal Medicines Using Pattern Recognition Techniques with Near-Infrared Reflectance Spectra. Microchemical Journal, 63, 61-70.
9. Savitzky, A. and Golay, M.J.E. (1964) Smoothing and Differentiation of Data by Simplified Least-Squares Procedures. Analytical Chemistry, 36, 1627-1639.
10. Barnes, R.J., Dhanoa, M.S. and Lister, S.J. (1989) Standard Normal Variate Transformation and De-Trending of Near- Infrared Diffuse Reflectance Spectra. Applied Spectroscopy, 43, 772-777.
11. Davies, A.M.C. and Fearn, T. (2005) Back to Basics: The Principles of Principal Component Analysis. Spectroscopy Europe, Tony Davies Column, 16, 20-23.
12. Næs, T., Isaksson, T., Fearn, T. and Davies, T. (2002) A User-Friendly Guide to Multivariate Calibration and Classification. NIR Publications, Chichester.
13. Barker, M. and Rayens, W. (2003) Partial Least Squares for Discrimination. Journal of Chemometrics, 17, 166-173.
14. Zhu, Y., Fearn, T., Samuel, D., Dhar, A., Hameed, O., Bown, S.G. and Lovat, L.B. (2008) Error Removal by Orthogonal Subtraction (EROS): A Customised Pre-Treatment for Spectroscopic Data. Journal of Chemometrics, 22, 130- 134.

REFERENCES

15. Roger, J.M., Palagos, B., Guillaume, S. and Bellon-Maurel, V. (2005) Discriminating from Highly Multivariate Data by Focal Eigen Function Discriminant Analysis; Application to NIR Spectra. Chemometrics and Intelligent Laboratory Systems, 79, 31-41.
16. Lelong, C.C.D., Roger, J.M., Brégand, S., Dubertret, F., Lanore, M., Sitorus, N.A., Raharjo, D.A. and Caliman, J.P. (2010) Evaluation of Oil-Palm Fungal Disease Infestation with Canopy Hyperspectral Reflectance Data. Sensors, 10, 734-747.
17. Yue, G.G.L., Fung, K.P., Tse, G.M.K., Leung, P.C. and Lau, C.B.S. (2006) Comparative Studies of Various Ganoderma Species and Their Different Parts with Regard to Their Antitumor and Immunomodulating Activities in Vitro. Journal of Alternative and Complementary Medicine, 12, 777-789.
18. R Core Team (2012) R: A Language and Environment for Statistical Computing. R Foundation for Statistical Computing, Vienna, Austria.
19. Chen, X.L., Liu, X.C., Sheng, D.P., Huang, D.K., Li, W.Z. and Wang, X. (2012) Distinction of Broken Cellular Wall Ganoderma lucidum Spores and G. lucidum Spores Using FTIR Microspectroscopy. Spectrochimica Acta Part A, 97, 667-672.
20. Gorgulu, S.T., Dogan, M. and Severcan, F. (2007) The Characterization and Differentiation of Higher Plants by Fourier Transform Infrared Spectroscopy. Applied Spectroscopy, 61, 300-308.
21. Elumba, Z.S., Teves, F.G. and Madamba, M.R.S.B. (2013) DNA-Binding and Cytotoxic Activities of Supercritical- CO_2 Extracts of Ganoderma lucidum Collected from the Wild of Bukidnon Province, Philippines. International Research Journal of Biological Sciences, 2, 62-68.
22. Kovac-Besovic, E.E., Duric, K., Kalodera, Z. and Sofic, E. (2009) Identification and Isolation of Pharmacologically Active Triterpenes in Betulae Cortex, Betula Pendula Roth., Betulaceae. Bosnian Journal of Basic Medical Sciences, 9, 31-38.
23. Paterson, R.R.M. (2006) Ganoderma—A Therapeutic Fungal Biofactory. Phytochemistry, 67, 1985-2001.
24. Sliva, D. (2003) Ganoderma lucidum (Reishi) in Cancer Treatment. Integrative Cancer Therapies, 2, 358-364.
25. Kao, C.H.J., Jesuthasan, A.C., Bishop, K.S., Glucina, M.P. and Ferguson, L.R. (2013) Anti-Cancer Activities of Ganoderma lucidum: Active Ingredients and Pathways. Functional Foods in Health and Disease, 3, 48-65.
26. Joshi, D.D. (2012) Herbal Drugs and Fingerprints. Springer, India.
27. Sun, S.Q., Zhou, Q. and Chen, J.B. (2010) Analysis of Traditional Chinese Medicine by Infrared Spectroscopy (in Chinese). Chemical Industry Press, Beijing.

28. Bombalska, A., Mularczyk-Oliwa, M., Kwaśny, M., Włodarski, M., Kaliszewski, M., Kopczy?ski, K., Szpakowska, M. and Trafny, E.A. (2011) Classification of the Biological Material with Use of FTIR Spectroscopy and Statistical Analysis. Spectrochimica Acta Part A: Molecular and Biomolecular Spectroscopy, 78, 1221-1226.
29. Weng, C.J. and Yen, G.C. (2010) The in Vitro and in Vivo Experimental Evidences Disclose the Chemopreventive Effects of Ganoderma lucidum on Cancer Invasion and Metastasis. Clinical & Experimental Metastasis, 27, 361-369.

CHAPTER 7
Spectroscopic Discrimination of Bone Samples from Various Species

Gregory McLaughlin, Igor K. Lednev*

Department of Chemistry, University at Albany, State University of New York, Albany, USA

ABSTRACT

Determining the species of origin of skeletal remains is critical in a forensic and anthropologic context. However, there are very few methods that use a chemical approach to assist in this determination. In this study, Raman spectroscopy was used to discriminate bone samples originating from four different species (bovine, porcine, turkey and chicken). Spectra were obtained using a near infrared laser at 785-nm. All spectra were combined in a single matrix and processed using partial least squares discriminate analysis (PLS-DA) with leave-one-out cross-validation. Three components were found to adequately describe the system. The first two components which contributed over 85% of spectral data was seen to completely separate the four species of origin in a two dimensional scores plot. A 95% confidence interval was draw around score points of each species class with very slight overlap. The first two components were seen to have large contributions from bioapatite and collagen, the main components of bone. This study serves as a preliminary investigation to evaluate the effectiveness of Raman spectroscopy to discriminate the species of origin of bone tissue.

Keywords: Forensic Science; Bone Tissue; Raman Spectroscopy; Chemometrics

1. INTRODUCTION

Skeletal remains found at a suspected crime scene are of particular interest to forensic investigators. However, there is little forensic importance if the skeletal remains are in fact non-human. The critical determination of whether skeletal remains are human falls on the forensic anthropologist [1]. This is relatively simple task when analyzing a full skeleton or complete bone, but can be very difficult when only fragmentary remains are obtained.

Visual identification methods are well studied and accepted in the scientific community. So much is known about the skeletal features and morphology of human remains that the determination of human origin of a whole bone is considered relatively simple. However, these methods are complicated when only a fragment of bone is in question. Small bone fragments are common in mass disasters such as earth quakes, terrorist attacks, or military battles.

Bone fragments can be devoid of any species specific morphological characteristics. In these cases, a histological analysis is often employed [2]. This approach requires imbedding and cross sectioning of samples and is therefore a destructive process. This type of analysis is considerably destructive because of the processing involved to obtain a cross section of bone.

Metric analysis of Haversian canals in bones has been used with varying success in the verification of human remains. This type of analysis involves measurements the narrow channel in the center of the osteon substructure and comparing to known values and indexes. A recent study by Cattaneo et al. (2009) [3] has shown significant limits with this type of evaluation. Although they reported a success rate of 70% with adult long bones, this type of analysis was seen to be especially problematic in young human. Such samples had an error rates as high as 93% in the case of newborn long bones and 68% for newborn flat bones.

The field of forensic anthropology is highly specialized, making rapid identification practically impossible. There is an interest for a more rapid, high throughput, reliable and non-destructive technique to identify the species of origin of skeletal remains. A Raman spectroscopic approach offers several advantages. Foremost, this technique is non-destructive and highly flexible for other forensic identification tasks. If a robust, automated data treatment process is developed, an examiner can be minimally trained. This type of analysis could eliminate examiner bias because it is not dependent on subjective observations. Most importantly, using a microscopic Raman system, a spectrum can be obtained from an area in the scale of microns. While current methods are hindered when analyzing fragmentary remains, there would be no such impediment using this approach.

It is our hypothesis that a method for species discrimination based on chemical differences in bone can be established. Using Raman spectroscopy, it is possible that a non-destructive and rapid technique to distinguish human remains can provide a favorable alternative to the current methodology. A technique that avoids physical and chemical alterations is attractive in a forensic setting because evidence is suitable for further testing. The main objective of this study is to develop a method based on Raman spectroscopic data that can be used to discriminate species. The proposed methodology is potentially applicable to the fields of anthropology for cultural studies and forensic science for criminal investigations.

Raman spectroscopy is an extremely flexible analytical technique with a variety of forensic applications [4]. This technique is currently being explored for the identification and classification of drugs [5], gunshot residue [6], fibers [7], paint [8-10] and ink [11] samples. When a laser is applied to a sample, light

1. INTRODUCTION

will be scattered resulting in photons with an initial (elastic scattering) or changed energy. Inelastic scattering occurs when photons loose/gain energy to excite/deactivate molecular vibrations in the sample. A Raman spectrum is obtained by collecting and analyzing these inelastic photons. The intensity of the inelastic photons is plotted against the energy in wavenumbers. Chemical compounds will produce a unique and predictable spectrum that is a result of molecular vibrational modes. Raman spectroscopy is advantageous for the field of forensic science because of the limited sample preparation, rapid results and the in-field analysis possibilities with portable Raman instruments [12-14]. The laser light is typically non-destructive so further testing is not affected in any way.

Bone tissue is mainly composed of inorganic mineral imbedded in an organic protein matrix. The inorganic portion, bioapatite, is similar to the natural geologic mineral hydroxyapatite. Bioapatite is a carbonated form of this mineral. The organic portion consists mainly of type I collagen. By weight, the organic and inorganic portions of bone are roughly 30% and 70% respectively. The composition of bone tissue varies significantly between species. Aerssens et al. reports several measurable differences in cortical bone between human, pig, cow, dog, sheep, chicken, and rat [15]. Key interspecies bone composition differences noted were collagen content, dry ash weight and non-collagen protein content.

Virkler and Lednev have recently introduced a new approach based on the combination of Raman spectroscopy and advanced statistical analysis for identification of body fluid traces [4,16-21] for forensic purposes. The approach relies on the ability of Raman spectroscopy to probe and characterize intrinsically heterogeneous body fluid samples. Further research by Virkler and Lednev has led to the development of methodology to discriminate blood samples of different species [22] with the application of Raman spectroscopy. A recent publication in the Journal of Analytical Chemistry reports on this method of identification of species based on the Raman spectroscopic analysis of blood [22]. We have recently reported on potential application of Raman spectroscopy for determining the burial time based on bone remains [23]. Here we report on a preliminary investigation of species differentiation based on Raman spectroscopy of bones. Our hypothesis is based on the literature which demonstrates that vibrational spectroscopic analysis of bone components and mineralized tissue can discriminate species of origin [24-28]. Species discrimination using Raman spectroscopy of animal tusk has been previously investigated by several researchers. Animal tusk, or dentine, is a mineralized tissue very similar to bone. Dentine tends to be more mineralized and have less collagen than bone tissue. Shimoyama et al. (1997) reports that hard and soft mammoth dentine were discriminated using near infrared (NIR) FT-Raman spectroscopy [24]. Brody et al. (2001) report the discrimination of dentine from six mammalian species, with slight overlap, also using FT-Raman spectroscopy [26] and chemometeric processing. Edwards et al. report using a similar approach to discriminate between Asian and African elephant dentine in a forensic setting [28]. Discrimination in these studies is likely due to organic to inorganic ratio differences between species.

Considering the reported success of the studies of biological material conducted for the purposes of species discrimination, we intend to use similar methods for this study. The main purpose of this research is to prove the principle that bone can be spectroscopically discriminated according to species of origin. This has clear forensic applications, but could also be useful in the study of ancient cultures. For this preliminary investigation, various non-human species were used for practicality; a human bone was not requisite at this stage. If the developed methods prove successful, continued research will focus on the discrimination of human bone specimen from other animal bones.

2. EXPERIMENTAL PROCEDURE

2.1. Samples
Chicken, turkey, cow and pig bone samples were collected from a grocery meat market. Each bone was cleaned of muscle, fat and soft tissue before Raman spectroscopic measurements.

2.2. Raman Microscope
A Renishaw in Via Raman microscope with attached Leica microscope with 20X objective was used. The wavelength of the laser used was 785 nm with approximately 11.5 mW power. Data was collected using Wire 2.0 software and analyzed using Grams v7.01. A small area of cortical bone was cut and placed on the lower plate of a Nanonics AFM MultiView 1000 automatic mapping stage. The automatic mapping stage was used to sample an arbitrary square area of cortical bone measuring approximately 75 µm × 75 µm. Within this square area, 36 points (6 × 6 square distribution) were sampled. Scans were set to 35 seconds with five accumulations and with a full range of 3200 - 100 cm^{-1}.

2.3. Data Treatment
GRAMS v7.01 was used to remove cosmic rays and for baseline subtraction. The data was truncated between the range 3190 - 360 cm^{-1}. The spectral data was then imported as a single matrix into MATLAB version 7.9.0. MATLAB PLS_Toolbox 5.5 by Eigenvector Research Inc. was used for pretreatment and partial least squares discriminant analysis (PLS-DA). The data was normalized about the most intense peak (~960 cm^{-1}) and mean centered. The number of components was estimated using the root mean square error of the leave-one-out crossvalidation (RMSECV). Component analysis was performed using PLS-DA and a model was built based on the component loadings. MATLAB software was also used for spectral averaging.

3. RESULTS AND DISCUSSION

3.1. Main Approach

Previous Raman studies of bone tissue have been limited because of strong fluorescence interference [29]. Golcuk et al. (2006) used photobleaching, a potentially destructtive process, prior to acquiring bone Raman spectra at 532-nm excitation [30]. In this study, to reduce the fluorescence interference we utilized a 785-nm near infrared light for excitation. Two, 60 second accumulations provides high quality Raman spectra with the signal-to-noise ratio of 50:1 or better.

Bone tissue is expected to be heterogeneous, meaning that no individual Raman spectrum measured at a single spot could fully represent a specific tissue. For this reason, we utilized a previously developed a mapping method for acquiring Raman spectra from multiple spots for further statistical analysis. In this study, 36 individual spectra were obtained for each species surveyed using the automatic mapping technique. PLS-DA was used because of the high discriminatory potential of this technique [31,32].

3.2. Spectral Analysis

The spectra were averaged for visual comparison and are shown in **Figure 1**. The averaged spectra between turkey, chicken, and pig bone samples are visually very similar. The peak count and position are identical between these species. There are, however, relative peak intensity variations that could be indicative of the variation in the bone tissue composition for various species. For example, the relative intensities of 1075 cm^{-1} and 1042 cm^{-1} Raman peaks vary (shaded portion of **Figure 1**). The Amide I peak at 1660 cm^{-1}, which is attributed to the contribution from the collagen organic component [33], varies in intensity between samples. The intensity a small peak at 1003 cm^{-1} also changes in intensity between these spectra. This peak is assigned to the phenylalanine amino acid breathing mode.

The averaged spectrum from cow is most distinct from the other three as it appears to be missing several peaks, including three peaks in the 800 - 900 cm^{-1} range, attributed to δ CCH at 855 cm^{-1}(collagen), v C-C 875 cm^{-1} (proline) and v C-C 920 cm^{-1} (collagen) vibrational modes [34]. The peak pattern in the CH$_2$ region at 2900 cm^{-1} for cow is comparably less intense than other species sampled. This is also seen in the Amide I, Amide III and δ NH bands. Excluding this irregularity, the Raman spectra between species are very similar.

3.3. Determining the Number of Components

Leave-one-out cross-validation was performed and the root mean square error of cross-validation (RMSECV) versus the component plot is shown in **Figure 2**. In this plot, a local minimum or a distinct change in slope will occur at the number of components which describe the system. There are such local minimums around n = 2 - 3. Between the four classes, either two or three components should adequately describe the system. It was seen that the first two components, which contributed over 85% of spectral data, were best at discriminating classes and other components did not vary much between classes. For the purposes of this study, a model built using the first two components is therefore sufficient.

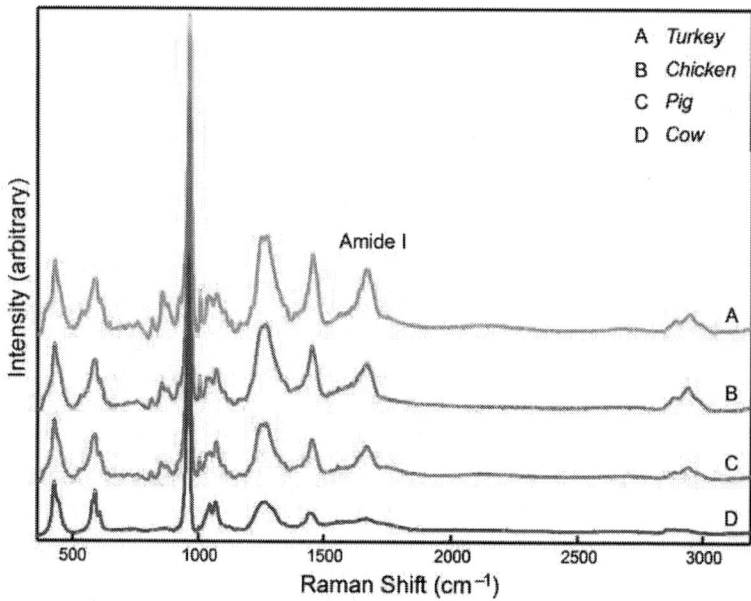

Figure 1. A stack plot showing the combined accumulations of Raman spectra for cow, pig, turkey and chicken. The shaded region is highlights peaks at 1075 and 1042 cm–1 which is discussed in Section 3.2. Raman spectra are baseline corrected to aid comparison.

3.4. Component Analysis

PCA is a multivariate data analysis method. This analysis describes a large data set with a small number of extracted components. These components have high variation which serves to highlight patterns in the dataset. This allows relationships in data to be efficiently described and represented by a few components. PLS-DA is an extension of PCA which accounts for class definitions and maximizes the separation of classes. In this case, the classes are the species identity i.e. cow, chicken, pig and turkey.

Component analysis is almost completely automated using MATLAB software and the Eigenvector PLS toolbox. First, the accumulated spectral data is combined into a single matrix. Normalization and mean centering preprocessing is performed on the data matrix to prepare the data for component analysis. When the PLS-DA algorithm is applied to this matrix, a number of components are extracted. The first component accounts for the majority of data variation within the matrix and is usually the most significant for discriminatory purposes. The following components progressively account for less data

3. RESULTS AND DISCUSSION

variance. The shape of a component is called its loading. Next, each individual spectrum is superimposed against the loading of a component to obtain a score. The more a given spectrum has features common to a component, the higher the score. A scores plot is usually displayed in two or three dimensions to graphically display the scores on multiple components. For example, a scores plot in two dimensions usually displays the scores on component one on the x-axis and the scores on component two on the yaxis. The actual numerical score of a given point is arbitrary because the zero, zero (0,0) point in a two dimensional scores plot is positioned at the total average score.

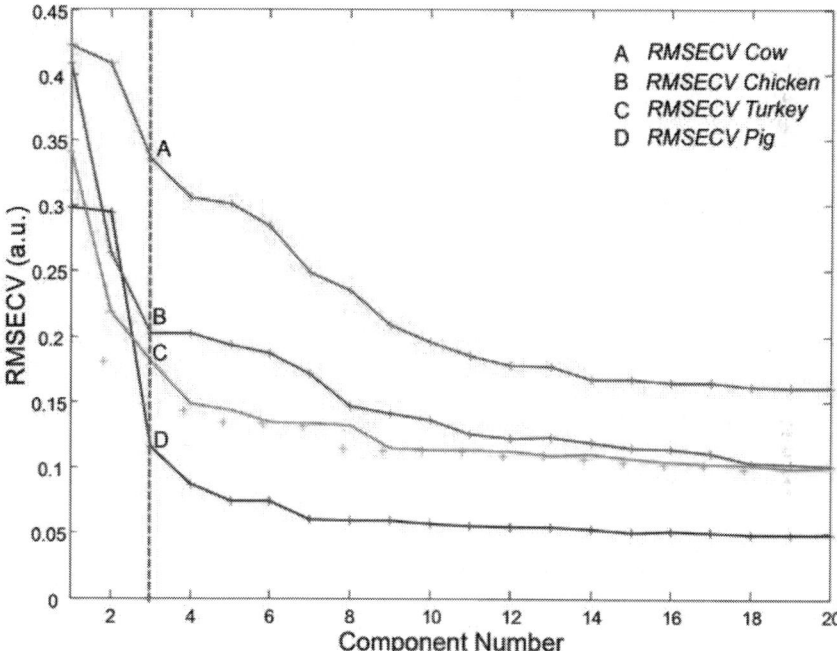

Figure 2. A root mean square error of cross-validation (RMSECV) plot versus the number of components. The break at n = 3 indicates that there are three components.

PLS-DA is very useful for discriminating samples that have similar chemical constituents. In the case of bone tissue, chemical constituents are essentially identical between species. All bone tissue is expected to be comprised of mostly bioapatite and type I collagen. However, the percent composition is expected to vary between species. Since the main components of bone, bioapatite and collagen, are detectable with Raman spectroscopy, PLS-DA analysis is appropriate for discriminatory purposes.

The components derived from this type of data analysis can often be chemically informative [35]. It is seen in this study that the components loadings reflect the spectral signatures of collagen and bioapatite. Raman signals from these materials have been previously separated and identified in a modern mouse bone. Freeman et al. (2002) reports the Raman spectrum from both organic and inorganic constituents [29]. A general rule is that the mineral phase of bone has a majority of Raman bands below 1000 cm^{-1} with a distinct sharp peak centered at 960 cm^{-1}. Conversely, the organic phase contributes a majority of peaks towards higher wavenumbers than 1000 cm^{-1} and are generally broad. Component assignments were made by comparing loading patterns with previously published Raman spectrum of isolated bone constituents.

The loadings for components one and two are shown in **Figure 3**. For component one, there are major positive features in the 3000 - 2800 cm^{-1} range as well as a triplet which has peaks centered at around 1260 cm^{-1}, 1450 cm^{-1} and 1670 cm^{-1}. This positive triplet corresponds to the Amide III, δ NH and Amide I Raman bands respectively. Hence, this component is dominated by organic contributions.

Conversely, component two is dominated by inorganic features. The dominant peak centered at 960 cm^{-1} is the v_1 PO_4^{3-} vibrational mode. The two peaks at lower wavenumbers from this, centered around 430 cm^{-1} and 590 cm^{-1} correspond to v_2 PO_4^{3-} and v_4 PO_4^{3-} vibrational modes. The peak at 1070 cm^{-1} corresponds to the carbonate ion (v_3 CO_3^{2-}) vibrational mode. The broad hump centered around 1260 cm^{-1} corresponds to the Amide III vibrational mode.

These component loading features are in agreement with isolated spectra of collagen and bioapatite published by Freeman et al. (2002) Although there is slight mutual contribution, in general component one represents a collagen contribution, while component two is representative of bioapatite.

The first two components were used to create a scores plot to determine if spectra will group with respect to species of origin. A two dimensional component scores plot is shown in **Figure 4**. A 95% confidence interval ellipsoid is drawn around each class. It is apparent that the classes are completely separated with respect to species of origin and there is only slight group overlap of the ellipsoid representing a 95% confidence interval.

3. RESULTS AND DISCUSSION

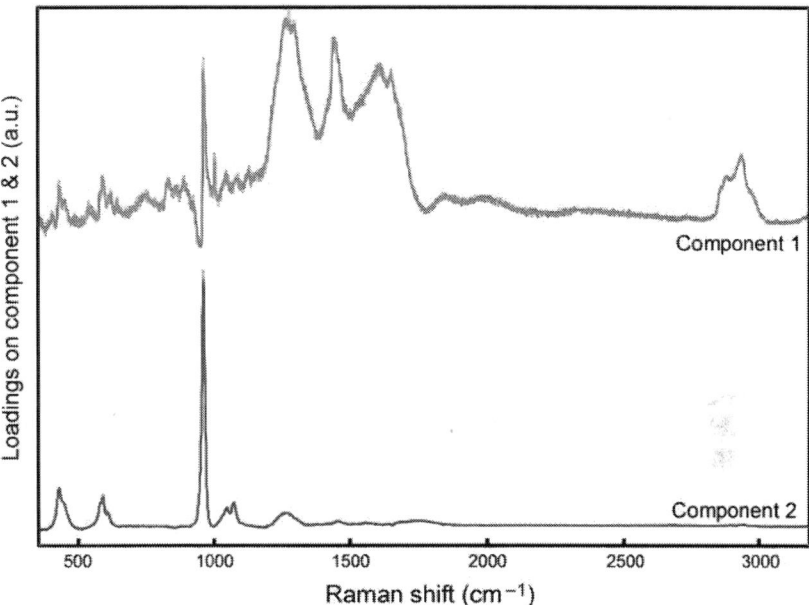

Figure 3. The component loading for spectral component one (top/red) and two (bottom/blue).

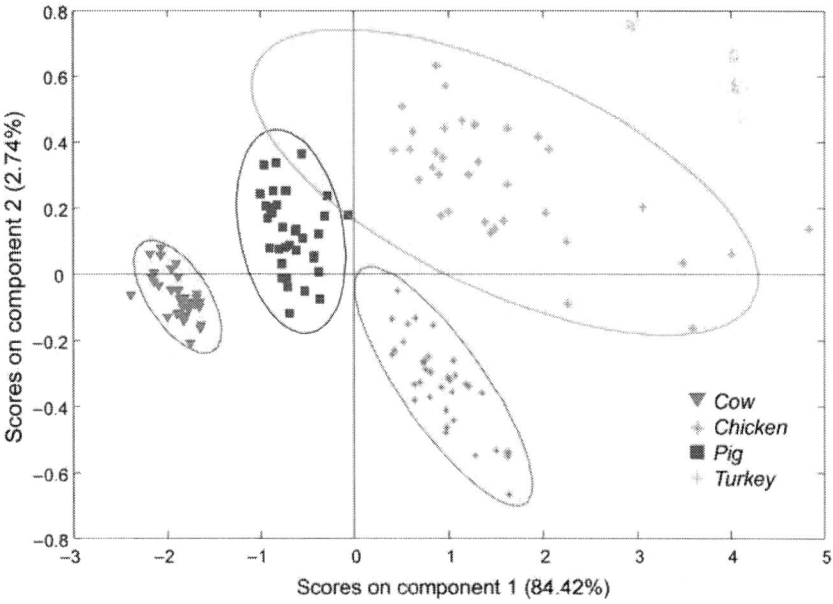

Figure 4. A two dimensional scores plot for cow, chicken, turkey and pig based on the first two spectral components. A 95% confidence ellipsoid encompasses each species class.

4. CONCLUSION

The goal of this research project was to investigate the potential of Raman spectroscopy to discriminate between bone samples based on species of origin. Previous research demonstrates that the chemical composition of bone is measurably different in various mammalian species [15]. This approach is appropriate since components of bone are easily detected using Raman spectroscopy. PLS-DA was chosen for this purpose because of its powerful discrimination ability. Using this method, spectra from four species were successfully discriminated with no overlap. The model built for discrimination reflects the chemical composition of bone material. The separation of groups is likely due to the differences in bone composition that have been previously reported in different species. These positive results demonstrate that the species of origin determination of bone in a practical setting may be possible. This would be a substantial contribution to forensic anthropology, a field that is currently lacking in spectroscopic analysis.

Further research is necessary for a more complete representation of each species. Variables such as specimen age, bone type, and gender are known to affect bone composition and need to be controlled. In addition, degenerative factors that change the composition of bone after death need to be analyzed for this approach to have practical applications. To prove forensic applicability, a human bone also needs to be sampled.

REFERENCES

1. M. Houck and J. Siegel, "Fundamentals of Forensic Science," Academic Press, San Diego, 2006.
2. A. Chamberlain, "Human Remains," University of California Press, Los Angeles, 1994.
3. C. Cattaneo, D. Porta, D. Gibelli and C. Gamba, "Histological Determination of the Human Origin of Bone Fragments," Journal of Forensic Sciences, Vol. 54, No. 3, 2009, pp. 531-533.
4. K. Virkler and I. K. Lednev, "Analysis of Body Fluids for Forensic Purposes: From Laboratory Testing to Non-destructive Rapid Confirmatory Identification at a Crime Scene," Forensic Science International, Vol. 188, No. 1-3, 2009, pp. 1-17.
5. C. M. Hodges and J. Akhavan, "The Use of FourierTransform Raman-Spectroscopy in the Forensic Identification of Illicit Drugs and Explosives," Spectrochimica Acta Part A—Molecular and Biomolecular Spectroscopy, Vol. 46, 1990, pp. 303-307.
6. E. M. A. Ali, H. G. M. Edwards and I. J. Scowen, "In-situ Detection of Single Particles of Explosive on Clothing with Confocal Raman Microscopy," Talanta, Vol. 78, 2009, pp. 1201-1203.

7. G. Jochem and R. J. Lehnert, "On the Potential of Raman Microscopy for the Forensic Analysis of Coloured Textile Fibres," Science & Justice, Vol. 42, No. 2, 2002, pp. 215-221.
8. J. Zieba-Palus and R. Borusiewicz, "Examination of Multilayer Paint Coats by the use of Infrared, Raman and XRF Spectroscopy for Forensic Purposes," Journal of Molecular Structure, Vol. 792-793, 2006, pp. 286-292.
9. S. E. J. Bell, L. A. Fido, S. J. Speers and J. Armstrong, "Raman Spectroscopy for Rapid, One-step Forensic Analysis of Paint," Forensic Science International, Vol. 136, 2003, pp. 354-355.
10. S. E. J. Bell, L. A. Fido, S. J. Speers, W. J. Armstrong and S. Spratt, "Forensic Analysis of Architectural Finishes using Fourier Transform Infrared and Raman Spectroscopy, Part II: White Paint," Applied Spectroscopy, Vol. 59, No. 11, 2005, pp. 1340-1346.
11. J. Mania, B. Trzcinska, M. Kunicki and P. Kooecielniak, "Comparison of the CE method with FTIR and Raman Spectrometry in the Field of Forensic Ink Analysis," Forensic Science International, Vol. 136, 2003, pp. 75-75.
12. B. Eckenrode and E. Bartick, "Portable Raman Spectroscopy Systems for Field Analysis," Forensic Science Communications, Vol. 3, 2001.
13. M. D. Hargreaves, K. Page, T. Munshi, R. Tomsett, G. Lynch and H. G. M. Edwards, "Analysis of Seized Drugs using Portable Raman Spectroscopy in an Airport Environment—A Proof of Principle Study," Journal of Raman Spectroscopy, Vol. 39, No. 7, 2008, pp. 873-880.
14. M. Perez-Alonso, K. Castro, I. Martinez-Arkarazo, M. Angulo, M. A. Olazabal and J. M. Madariaga, "Analysis of Bulk and Inorganic Degradation Products of Stones, Mortars and Wall Paintings by Portable Raman Microprobe Spectroscopy," Analytical and Bioanalytical Chemistry, Vol. 379, No. 1, 2004, pp. 42-50.
15. J. Aerssens, S. Boonen, G. Lowet and J. Dequeker, "Interspecies Differences in Bone Composition, Density, and Quality: Potential Implications for In Vivo Bone Research," Endocrinology, Vol. 139, No. 2, 1998, pp. 663- 670.
16. K. Virkler and I. K. Lednev, "Raman Spectroscopy Offers Great Potential for the Nondestructive Confirmatory Identification of Body Fluids," Forensic Science International, Vol. 181, No. 1-3, 2008, pp. e1-e5.
17. K. Virkler and I. K. Lednev, "Raman Spectroscopic Signature of Semen and its Potential Application to Forensic Body Fluid Identification," Forensic Science International, Vol. 193, No. 1-3, 2009, pp. 56-62.
18. K. Virkler and I. K. Lednev, "Raman Spectroscopic Signature of Blood and its Potential Application to Forensic Body Fluid Identification," Analytical and Bioanalytical Chemistry, Vol. 396, No. 1, 2010, pp. 525-534.

19. K. Virkler and I. K. Lednev, "Forensic Body Fluid Identification: The Raman Spectroscopic Signature of Saliva," Analyst, Vol. 135, 2010, pp. 512-517.
20. V. Sikirzhytski, K. Virkler and I. K. Lednev, "Discriminant Analysis of Raman Spectra for Body Fluid Identification for Forensic Purposes," Sensors, Vol. 10, No. 4, 2010, pp. 2869-2884.
21. A. Sikirzhytskaya, V. Sikirzhytski and I. K. Lednev, "Raman Spectroscopic Signature of Vaginal Fluid and its Potential Application in Forensic Body Fluid Identification," Forensic Science International, 2011, Article in Press.
22. K. Virkler and I. Lednev, "Blood Species Identification for Forensic Purposes Using Raman Spectroscopy Combined with Advanced Statistical Analysis," Analytical Chemistry, Vol. 81, No. 18, 2009, pp. 7773-7777.
23. G. McLaughlin and I. K. Lednev, "Potential Application of Raman Spectroscopy for Determining Burial Duration of Skeletal Remains" Analytical and Bioanalytical Chemistry, Vol. 401, No. 8, 2011, pp. 2511-2518.
24. M. Shimoyama, H. Maeda, H. Sato, T. Ninomiya and Y. Ozaki, "Nondestructive Discrimination of Biological Materials by Near-infrared Fourier Transform Raman Spectroscopy and Chemometrics: Discrimination Among Hard and Soft Ivories of African Elephants and Mammoth Tusks and Prediction of Specific Gravity of the Ivories," Applied Spectroscopy, Vol. 51, No. 8, 1997, pp. 1154-1158.
25. M. Shimoyama, S. Morimoto, and Y. Ozaki, "Non-destructive Analysis of the Two Subspecies of African Elephants, Mammoth, Hippopotamus, and Sperm Whale Ivories by Visible and Short-wave Near Infrared Spectroscopy and Chemometrics," Analyst, Vol. 129, 2004, pp. 559-563.
26. R. H. Brody, H. G. M. Edwards, and A. M. Pollard, "Chemometric Methods Applied to the Differentiation of Fourier-Transform Raman Spectra of Ivories," Analytica Chimica Acta, Vol. 427, 2001, pp. 223-232.
27. D. M. Hashim, Y. B. C. Man, R. Norakasha, M. Shuhaimi, Y. Salmah, and Z. A. Syahariza, "Potential Use of Fourier Transform Infrared Spectroscopy for Differentiation of Bovine and Porcine Gelatins," Food Chemistry, Vol. 118, No. 3, 2009, pp. 856-860.
28. H. G. M. Edwards, N. F. N. Hassan and N. Arya, "Evaluation of Raman Spectroscopy and Application of Chemometric Methods for the Differentiation of Contemporary Ivory Specimens I: Elephant and Mammalian Species," Journal of Raman Spectroscopy, Vol. 37, No. 1-3, 2006, pp. 353-360.

29. J. J. Freeman and M. J. Silva, "Separation of the Raman Spectral Signatures of Bioapatite and Collagen in Compact Mouse Bone Bleached with Hydrogen Peroxide," Applied Spectroscopy, Vol. 56, No. 6, 2002, pp. 770-775.
30. K. Golcuk, G. S. Mandair, A. F. Callender, N. Sahar, D. H. Kohn and M. D. Morris, "Is Photobleaching Necessary for Raman Imaging of Bone Tissue using a Green Laser?" Biochimica Et Biophysica Acta-Biomembranes, Vol. 1758, No. 7, 2006, pp. 868-873.
31. S. Lars and W. Svante, "Partial Least Squares Analysis with Cross-Validation for the Two-Class Problem: A Monte Carlo Study," Journal of Chemometrics, Vol. 1, 1987, pp. 185-196.
32. M. Barker and W. Rayens, "Partial Least Squares for Discrimination," Journal of Chemometrics, Vol. 17, No. 3, 2003, pp. 166-173.
33. R. Smith and I. Rehman, "Fourier-Transform RamanSpectroscopic Studies of Human Bone," Journal of Materials Science-Materials in Medicine, Vol. 5, 9-10, No. 1994, pp. 775-778.
34. A. Awonusi, M. D. Morris and M. M. Tecklenburg, "Carbonate Assignment and Calibration in the Raman Spectrum of Apatite," Calcified Tissue International, Vol. 81, No. 1, 2007, pp. 46-52.
35. H. Nocairi, E. M. Qannari, E. Vigneau and D. Bertrand, "Discrimination on Latent Components with Respect to Patterns. Application to Multicollinear Data," Computational Statistics & Data Analysis, Vol. 48, No. 1, 2005, pp. 139-147.

CHAPTER 8

New Approachs in Drug Quality Control: Matrices and Chemometrics

Sigrid Mennickent[1], M. de Diego[1], B. Schulz[1], M. Vega[2] and C. G. Godoy[1]

[1] Department of Pharmacy, Faculty of Pharmacy, University of Concepción, Concepción,, Chile
[2] Department of Bromatology, Nutrition and Dietetic, Faculty of Pharmacy, University of Concepción, Concepción, Chile

1. QUALITY CONTROL

Quality control refers to the process of quality evaluation that focuses on the internal measurement of the quality of a process, institution, product, service, or other. Often used interchangeably with *quality management* and *quality assurance* [1-3].

2. DRUG QUALITY CONTROL

Quality Assurance plays a very important role in making sure that the GMP standards are met and products comply with the international quality standards. The main functions carried out by drug quality control are:

- Approval of raw materials
- Monitoring of manufacturing processes
- Approval of finished products
- Documentation of technical information
- Implementation of cGMP

Manufacturing processes are monitored and controlled by testing of raw materials, in-process parameters. Final active pharmaceutical ingredients and dosage forms are tested for specified parameters before release. Analytical testing is carried out with highly sophisticated instruments: viz. HPLC, GC, IR, UV spectrophotometer mettler titrators, particle size analyzer etc.

All the analytical test procedures and manufacturing procedures are well documented and revision is undertaken as per specified protocol. Analytical methods are validated to give the reproducible results. Stability study as per stability protocol is considered to be very important area of Quality Assurance.

Automated systems are becoming increasingly important tools for appropriate monitoring and controlling of the pharmaceutical packaging process. Solutions for comprehensive quality assurance or production data acquisition and evaluation are just as important as applications that meet the legislative requirements of different countries in terms of serial numbering and the unique marking of products.

Quality control involves many phases, such as sample collection, measuring, analysis of results, and the approval/rejection of the batch. Nonetheless, the most important thing is the continuity and systematization of the quality control.

Effective process validation contributes significantly to assuring drug quality. The basic principle of quality assurance is that a drug should be produced that is fit for its intended use. This principle incorporates the understanding that the following conditions exist: Quality, safety, and efficacy are designed or built into the product.

Quality cannot be adequately assured merely by in-process and finished-product inspection or testing.

Process validation is defined as the collection and evaluation of data, from the process design stage through commercial production, which establishes scientific evidence that a process is capable of consistently delivering quality product. Process validation involves a series of activities taking place over the lifecycle of the product and process. Usually, process validation includes three stages:

- Process Design: The commercial manufacturing process is defined during this stage based on knowledge gained through development and scale-up activities.
- Process Qualification: During this stage, the process design is evaluated to determine if the process is capable of reproducible commercial manufacturing.
- Process Verification: Ongoing assurance is gained during routine production that the process remains in a state of control [1-3].

3. NEW APPROACHES IN DRUG QUALITY CONTROL

3.1. Matrices
3.1.1. Residue analysis of pharmaceuticals in the aquatic environment
Residue analysis of pharmaceuticals in the aquatic environment has attracted considerable interest during the last few years.

Traces of such compounds have been detected in surface water samples from all countries where pharmaceuticals are widely in use.

Pharmaceutically active compounds have captured the attention of the scientific community because such pollutants result not primarily from manufacturing but from widespread, continual use in human and veterinary clinical practice. The biological activity of these compounds can lead to adverse effects in aquatic ecosystems and potentially have an impact on drinking-water supplies [4].

In the human body, pharmaceuticals can be transformed to one or more metabolites and excreted as a mixture of parent compound and metabolites, in which the parent compound is often the minor component. However, some drugs are poorly metabolized and are excreted unchanged. The degree of metabolism depends on a number of parameters, including age, gender and ethnicity, the constitution of the patient and the time of administration. Drug-drug interactions caused by enzyme induction or inhibition, as well as enhanced metabolism due to previous exposure, can also influence the pharmacokinetics of drugs [5].

Both the parent compound and the metabolites enter the aquatic environment once they are excreted from the human body. Monitoring studies in the environment have demonstrated the discharge of pharmaceuticals and their metabolites through municipal wastewater-treatment plants (WWTPs). Although unchanged drugs can undergo biochemical transformations during sewage treatment, some studies indicate that the absence of pharmaceutical compounds in treated water does not necessarily imply their complete removal. In most instances, human drugs are metabolized in the body to more polar compounds that are more likely to pass through the WWTP. In some cases, pharmaceuticals and their human metabolites can be microbially degraded in the activated sludge treatment.

Knowledge of the formation of stable metabolites in WWTPs is also important in order to understand the environmental fate of the parent compound. Once in the environment, these compounds can be transported and distributed in rivers, streams, and possibly further biodegraded. For most pharmaceuticals and their biotransformation products, these pathways in the aquatic environment are largely unknown, and investigations into their occurrence in environmental compartments are still rare.

Studies have been carried out to investigate their fate not only in surface waters, but also in sediment and soil environments. By nature, most pharmaceuticals are designed to be at least moderately water-soluble and to possess half-lives in the human body in the range of hours. Because human and microbial degradates will generally coexist with their parent compounds in the environment, indicators that summarize all the information on parent substances and degradates would be important instruments for decision-making and assessment [6].

Progress in instrumental analytical chemistry has resulted in the availability of methods that allow a monitoring of these pollutants at ng levels.

Improvements in detection limits over the past years have mainly been due to sophisticated mass spectrometric detection techniques. Furthermore, robust sample preparation and pre-concentration protocols have contributed significantly to the achievements observed so far.

Nowadays it is a well-established fact that pharmaceutical drugs used during medical treatment may partly be excreted in an un-metabolized form, enter municipal sewage systems, and can even survive the passage through the sewage treatment plant. Therefore, sewage treatment plant effluents are the major source for introduction of pharmaceuticals into the aquatic environment. Furthermore, pharmaceuticals employed in veterinary medicine may be introduced into soil (and eventually into water) via manure, or may find a direct way into the aquatic system when used in fish farms.

Unfortunately, the consequences of continuous presence of low concentrations of pharmaceuticals for the ecosystem are still not fully known.

In many cases, the analytical procedures for residue analysis of pharmaceutical drugs nowadays available includes a pre-concentration and clean-up step by solid-phase extraction or related techniques, followed by chromatography in combination with mass spectrometry (MS) as detector.

Although GC–MS may still be the perfect technique for certain classes of pharmaceuticals, high-performance liquid chromatography (HPLC) hyphenated with atmospheric pressure ionization-MS has established itself as the better choice for simultaneous determination of pharmaceuticals of widely differing structures.

The concentration levels of pharmaceuticals found in environmental water samples are generally too low to allow a direct injection into a chromatographic system. Therefore, efficient pre-concentration steps are necessary which should also result in some sample clean-up. One of the most widely used sample treatment technique for residue analysis of pharmaceuticals in water is the extraction of the analytes by means of a solid sorbent.

This extraction procedure can be based on multiple equilibria between the liquid phase and the sorbent filled into a small cartridge (solid-phase extraction, SPE), or on a single equilibrium (sorptive extraction) [7-23].

3.1.1.1. Solid-Phase Extraction

Pharmaceuticals of adequate hydrophobicity can easily be pre-concentrated using any reversed-phase material such as alkyl-modified silica or polymer-based materials. Deprotonation of acidic compounds and protonation of basic compounds should be suppressed to ensure sufficient hydrophobicity of the analytes. Therefore, acidic pharmaceuticals should be pre-concentrated under acidic conditions, whereas basic analytes should be pre-concentrated at an alkaline pH. Alternatively, mixed-mode SPE materials can be used which exhibit both reversed-phase and cation-exchange properties due to the presence of sulfonic acid groups on the hydrophobic surface of the particles. Using acidified sample solutions, acidic and neutral analytes would be extracted by hydrophobic interactions, whereas protonated basic analytes would interact via ion exchange mechanisms.

A recent review has summarized new SPE materials that can improve the recoveries for polar analytes. These materials are mainly polymeric sorbents that improve the retention of polar compounds either by novel functional groups in the polymeric structure (resulting in a hydrophilic–hydrophobic balance material) or by considerably increased surface area. Some of these new materials have turned out to be well suited for multi-class analysis of pharmaceuticals in water samples. Nowadays, one of the most widely used sorbent is a copolymer of divinylbenzene and vinylpyrrolidone [7-23].

3.1.1.2. Sorptive Extraction

Sorptive extraction based on a single partitioning equilibrium of analytes between the aqueous sample and a solid sorbent includes solid-phase microextraction (SPME), stir-bar sorptive extraction (SBSE), and several related variants. Originally, these techniques were based on polydimethylsiloxane (PDMS) as material for trapping trace analytes from a water sample due to partitioning between the aqueous matrix and the PDMS phase. Besides PDMS, some alternative sorptive materials have become commercially available recently, such as polyacrylates, copolymers of PDMS with divinylbenzene, copolymers of polyethylene glycol with divinylbenzene, and mixtures of carboxen (an inorganic adsorbent) with PDMS or divinylbenzene [7-23].

3.1.1.3. Sample Pre-Concentration Procedures For Sediment And Sludge Samples

Extraction of pharmaceuticals from sediment and sludge is generally done by blending the sample with an organic solvent or with mixtures of aqueous buffers and organic solvents.

Ultrasonication is frequently applied to assist the extraction process.

Additional clean-up steps for the extract may be necessary employing SPE or liquid–liquid extraction. Somewhat more advanced procedures are based on pressurized liquid extraction (accelerated solvent extraction) which may need less time and less solvent consumption [7-23].

3.1.1.4. Derivatization Of The Compounds

Various groups of pharmaceuticals can be derivatized to make them suited for GC analysis. Typical derivatization reagents for acidic pharmaceuticals include pentafluorobenzylbromide, methyl chloromethanoate, methanol/BF3, or tetrabutylammonium salts (for derivatization during injection). Phenazone-type drugs have been derivatized by silylation using *N-tert*-butyldimethylsilyl-*N*-methyltrifluoroacetamide (MTBSTFA). Silylation procedures are also commonly used for synthetic estrogens [7-23].

3.2.1. Some Latest Researches In This Area

3.2.1.1. Pharmaceuticals In The Aquatic Environment: A Critical Review Of The Evidence For Health Effects In Fish

The authors review the current data on the presence and reported biological effects in fish of some of the most commonly detected pharmaceuticals in the aquatic environment; namely nonsteroidal anti-inflammatory drugs (NSAIDs), fibrates, beta-blockers, selective serotonin reuptake inhibitors (SSRIs), azoles, and antibiotics. Reported biological effects in fish in the laboratory have often been shown to be in accordance with known effects of pharmaceuticals in mammals. Water concentrations at which such effects have been reported, however, are generally, between microg L(-1) and mg L(-1), typically at least 1 order of magnitude higher than concentrations normally found in surface waters (ng L(-1)). There are exceptions to this, however, as for the case of synthetic oestrogens, which can induce biological effects in the low ng L(-1) range.

Although generally effect levels for pharmaceuticals are higher than those found in the environment, the risks to wild fish populations have not been thoroughly characterised, and there has been a lack of consideration given to the likely chronic nature of the exposures, or the potential for mixture effects. As global consumption of pharmaceuticals rises, an inevitable consequence is an increased level of contamination of surface and ground waters with these biologically active drugs, and thus in turn a greater potential for adverse effects in aquatic wildlife [24].

3.1.1.2. Human Pharmaceuticals, Hormones And Fragrances: The Challenge Of Micropollutants In Urban Water Management
The observed concentrations of pharmaceuticals and personal care products (PPCPs) in raw wastewater confirm that municipal wastewater represents the main disposal pathway for the PPCPs consumed in households, hospitals and industry. In sewage treatment plant effluents most PPCPs are still present, since many of these polar and persistent compounds are being removed only partially or, in some cases, not at all. Treated wastewater therefore represents an important point source for PPCPs into the environment. After passing a sewage treatment plant the treated wastewater is mostly discharged into rivers and streams or sometimes used to irrigate fields. If drinking water is produced using resources containing a substantial proportion of treated wastewater (e.g. from river water downstream of communities) the water cycle is closed and indirect potable reuse occurs. Human Pharmaceuticals, Hormones and Fragrances provides an overview of the occurrence, analytics, removal and environmental risk of pharmaceuticals and personal care products in wastewater, surface water and drinking water. [25].

3.2.1.2. Factors affecting the concentrations of pharmaceuticals released to the aquatic environment
Although recent research has demonstrated that pharmaceuticals are widely distributed in the aquatic environment, it is difficult to assess the threat that they pose to drinking water supplies or their rate of attenuation in natural systems without an adequate understanding of the sources of contamination. To identify pharmaceutical compounds of significance to water supplies in the United States, the authors have reviewed available data on the use of prescription drugs. Results of our analysis indicate that approximately 40 compounds could be present in municipal wastewater effluent at concentrations above 1,000 ng/L and at least 120 compounds could be present at concentrations above 1 ng/L. Important classes of prescription drugs include analgesics, beta-blockers, and antibiotics. Analysis of a group of the most commonly used pharmaceuticals in the United States indicates that they are ubiquitous in wastewater effluents. Authors have detected concentrations ranging from approximately 10- 3,000 ng/L for high use pharmaceuticals such as betablockers (*e.g.,* metoprolol, propranolol) and acidic drugs (*e.g.,* gemfibrozil, ibuprofen). The concentration of pharmaceuticals in effluent from conventional wastewater treatment plants is similar. Advanced wastewater treatment plants equipped with reverse osmosis systems reduce concentrations of pharmaceuticals below detection limits. In addition to removal during biological wastewater treatment, pharmaceuticals

also are attenuated in engineered natural systems (*i.e.,* treatment wetlands, ground water infiltration basins). Preliminary evidence suggests limited removal of pharmaceuticals in engineered treatment wetlands and nearly complete removal of pharmaceuticals during ground water infiltration [26].

3.2.1.3. A preliminary ecotoxicity study of pharmaceuticals in the marine environment

Environmental fates and effects of pharmaceuticals in the aquatic environment have been the focus of recent research in environmental ecotoxicology. Worldwide studies of common over-the-counter pharmaceuticals have reported detectable levels in the aquatic environment, but there are few studies examining impacts on marine habitats. These drugs can affect the functions of various vertebrates and invertebrates. The stability of two pharmaceuticals, cyclizine (CYC) and prochlorperazine (PCZ), in seawater was examined under light and dark conditions, as well as the toxicity of these compounds to larvae of the barnacle Balanus amphitrite, which is a cosmopolitan marine organism found in most of the world's oceans. CYC was very stable under all the tested conditions. On the other hand, PCZ degraded in light but not in the dark, and was more stable in seawater than fresh water. For the barnacle larvae, the LC50 of prochlorperazine was 0.93 microg/mL and the LC50 for CYC was approximately 0.04 microg/mL [27].

3.2.1.4. Estrogenic activity of pharmaceuticals in the aquatic environment

In the last years pharmaceuticals have aroused great interest as environmental pollutants for their toxic effects towards non target organisms. This study wants to draw attention to a further adverse effect of drugs, the endocrine interference. The most representative drugs of the widespread classes in environment were investigated. The YES-test and the E-screen assay were performed to detect the capability of these substances to bind the human estrogenic receptor alpha (hER alpha) in comparison with 17beta-estradiol. Out of 14 tested pharmaceuticals, 9 were positive to YES-assay and 11 were positive to E-screen assay; in particular, Furosemide and the fibrates (Bezafibrate, Fenofibrate and Gemfibrozil) gave the maximal estrogenic response. Tamoxifen showed its dual activity as agonist and antagonist of hER alpha [28].

3.2.1.5. Colloids as a sink for certain pharmaceuticals in the aquatic environment

The occurrence and fate of pharmaceuticals in the aquatic environment is recognized as one of the emerging issues in environmental chemistry and as a matter of public concern. Existing data tend to focus on the concentrations of pharmaceuticals in the aqueous phase, with limited studies on their concentrations in particulate phase such as sediments. Furthermore, current water quality monitoring does not differentiate between soluble and colloidal phases in water samples, hindering our understanding of the bioavailability and bioaccumulation of pharmaceuticals in aquatic organisms. In this study, an investigation was conducted into the concentrations and phase association (soluble, colloidal, suspended particulate matter or SPM) of selected pharmaceuticals (propranolol, sulfamethoxazole, meberverine, thioridazine, carbamazepine, tamoxifen, indomethacine, diclofenac, and meclofenamic acid) in river water, effluents from sewage treatment works (STW), and groundwater in the UK. Colloids were isolated by cross-flow ultrafiltration (CFUF). Water samples were extracted by solid-phase extraction (SPE), while SPM was extracted by microwave. All sample extracts were analyzed by liquid chromatography-tandem mass spectrometry (LC-MS/MS) in the multiple reaction monitoring.

Five compounds propranolol, sulfamethoxazole, carbamazepine, indomethacine, and diclofenac were detected in all samples, with carbamazepine showing the highest concentrations in all phases. The highest concentrations of these compounds were detected in STW effluents, confirming STW as a key source of these compounds in the aquatic environments. The calculation of partition coefficients of pharmaceuticals between SPM and filtrate, between SPM and soluble phase, and between colloids and soluble phase showed that intrinsic partition coefficients are between 25% and 96%, and between 18% and 82% higher than relevant observed partition coefficients values, and are much less variable. Secondly, K_{coc} values are 3–4 orders of magnitude greater than Kocint values, indicating that aquatic colloids are substantially more powerful sorbents for accumulating pharmaceuticals than sediments. Furthermore, mass balance calculations of pharmaceutical concentrations demonstrate that between 23% and 70% of propranolol, 17–62% of sulfamethoxazole, 7–58% of carbamazepine, 19–84% of indomethacine, and 9–74% of diclofenac are present in the colloidal phase.

The results provide direct evidence that sorption to colloids provides an important sink for the pharmaceuticals in the aquatic environment. Such strong pharmaceutical/colloid interactions may provide a long-term storage of pharmaceuticals, hence, increasing their persistence while reducing their bioavailability in the environment.

Recommendations and perspectives from this study:

Pharmaceutical compounds have been detected not only in the aqueous phase but also in suspended particles; it is important, therefore, to have a holistic approach in future environmental fate investigation of pharmaceuticals. For example, more research is needed to assess the storage and long-term record of pharmaceutical residues in aquatic sediments by which benthic organisms will be most affected. Aquatic colloids have been shown to account for the accumulation of major fractions of total pharmaceutical concentrations in the aquatic environment, demonstrating unequivocally the importance of aquatic colloids as a sink for such residues in the aquatic systems. As aquatic colloids are abundant, ubiquitous, and highly powerful sorbents, they are expected to influence the bioavailability and bioaccumulation of such chemicals by aquatic organisms. It is therefore critical for colloids to be incorporated into water quality models for prediction and risk assessment purposes [29].

4. CHEMOMETRICS

Chemometrics is the science of extracting information from chemical systems by data-driven means. It is a highly interfacial discipline, using methods frequently employed in core data-analytic disciplines such as multivariate statistics, applied mathematics, and computer science, in order to address problems in chemistry, biochemistry, medicine, biology and chemical engineering.

Chemometrics is applied to solve both descriptive and predictive problems in experimental life sciences, especially in chemistry. In descriptive applications, properties of chemical systems are modeled with the intent of learning the underlying relationships and structure of the system (i.e., model understanding and identification). In predictive applications, properties of chemical systems are modeled with the intent of predicting new properties or behavior of interest. In both cases, the datasets can be small but are often very large and highly complex, involving hundreds to thousands of variables, and hundreds to thousands of cases or observations.

Chemometric techniques are particularly heavily used in analytical chemistry and metabolomics, and the development of improved chemometric methods of analysis also continues to advance the state of the art in analytical instrumentation and methodology. It is an application driven discipline, and thus while the standard chemometric methodologies are very widely used industrially, academic groups are dedicated to the continued development of chemometric theory, method and application development [30-33].

REFERENCES

1. USP 1999 The United States Pharmacopeia/ The National Formulary (USP 24/NF 19), United States Pharmacopeial Convection, Inc., Rockville
2. ICH 2003 The Sixth ICH International Conference on Armonization of Technical Requirements for Registration of Pharmaceuticals for Human Use, Osaka,
3. W. W. Buchberger, 2007 Analytica Chimica Acta 593 129 139
4. O. Jones, J. N. Lester, N. Voulvoulis, 2005 Trends in Biotechonology 23 4 163 167
5. B. Lemmer, 1996 Chronopharmacology- Cellular and Biochemical Interactions. Birkhäuser, New York, USA
6. A. Boxall, K. Fenner, D. W. Kolpin, S. Maund, 2004 Environ. Sci. Technol. 38 369A
7. Z. Moldovan, 2006 Chemosphere 64 1808
8. V. L. Cunningham, M. Buzby, T. Hutchinson, F. Mastrocco, N. Parke, N. Roden, 2006 Env. Sci. Technol 40 3457
9. K. Fent, A. A. Weston, D. Caminada, 2006 Aquatic Toxicol 76 122
10. M. Crane, C. Watts, T. Boucard, 2006 Sci. Tot. Environ 367 23

11. M. D. Hernando, M. Mezcua, A. R. Fernandez-Alba, D. Barcelo, 2006 Talanta 69 334
12. E. Benito-Peña, A. I. Partal-Rodera, M. E. Leon-Gonzalez, M. C. Moreno-Bondi, 2006 Anal. Chim. Acta 556 415
13. N. Fontanals, R. M. Marce, F. Borrull, 2005 Trends Anal. Chem 24 394.
14. M. J. Gomez, M. Petrovic, A. R. Fernandez-Alba, D. Barcelo, 2006 J. Chromatogr. A 1114 224
15. M. Gros, M. Petrovic, D. Barcelo, 2006 Talanta 70 678
16. M. Petrovic, M. Gros, D. Barcelo, 2006 J. Chromatogr. A 1124 68
17. R. A. Trenholm, B. J. Vanderford, J. C. Holady, D. J. Rexing, S. A. Snyder, 2006 Chemosphere 65 1990
18. P. H. Roberts, P. Bersuder, 2006 J. Chromatogr. A 1134 143
19. M. Himmelsbach, W. Buchberger, C. Klampfl, 2006 Electrophoresis 27 1220
20. W. Seitz, W. H. Weber, J. Q. Jiang, B. J. Lloyd, M. Maier, D. Maier, W. Schulz, 2006 Chemosphere 64 1318
21. O. J. Pozo, C. Guerrero, J. V. Sancho, M. Ibañez, E. Pitarch, E. Hogendoorn, F. Hernandez, 2006 J. Chromatogr. A 1103 83
22. S. Rodriguez-Mozaz, D. Lopez de Alda, D. Barcelo, 2007 J. Chromatogr. A 1152 97
23. J. B. Quintana, M. Miro, J. M. Estela, V. Cerda, 2006 Anal. Chem. 78 2832
24. J. Corcoran, M. J. Winter, C. R. Tyler, 2010 Crit, Rev. Toxicol 40 4 287 304
25. T. Ternes, 2006 Human Pharmaceuticals, Hormones and Fragrances: The Challenge of Micropollutants in Urban Water Management. Iwa Publishing
26. D. Sedlak, K. Pinkston, 2011 Factors affecting the concentrations of pharmaceuticals released to the aquatic environment University of California
27. A. M. Choong, S. L. Teo, J. L. Leow, P. C. Ho, 2006 J. Toxicol. Environ. Health A. 69 21 1959 1970
28. M. Isidori, M. Bellota, M. Cangiano, A. Parrella, 2009 Environ. Int 35 5 826 829
29. K. Maskaoui, J. Zhou, 2010 Environmental Science and Pollution Research 17 4 898 907
30. S. S. Verenitch, C. J. Lowe, A. Mazumder, 2006 J. Chromatogr. A 1116
31. P. J. Gemperline, (ed) 2006 Practical guide to chemometrics2nd Edition, CRC Press 10.1201/9781420018301
32. H. Mark, J. Workman, 2007 Chemometrics in spectroscopyAcademic Press-Elsevier
33. M. Maeder, Y. M. Neuhold, 2007 Practical Data Analysis in Chemistry Elsevier

CHAPTER 9
Application of Chemometrics to the Interpretation of Analytical Separations Data

James J. Harynuk, A. Paulina de la Mata and Nikolai A. Sinkov

Department of Chemistry, University of Alberta Canada

1. INTRODUCTION

Interesting real-world samples are almost always present as mixtures containing the analyte(s) of interest and a matrix of components that are irrelevant to answering the analytical question at hand. Additionally, the compounds comprising the matrix are usually present in far greater abundance (both number and concentration) than the analytes of interest, making quantification or even detection of these analytes difficult if not impossible.

When tasked with these types of samples, analysts turn to some form of separations technique such as gas or liquid chromatography (GC or LC) or capillary electrophoresis (CE) so that individual components in each sample may be quantified. More recently, more complex analytical questions are being probed, for example profiling blood or urine to identify a disease state or ascertaining the geographic origin of a food/beverage sample. These tasks often go beyond the simple quantification of one or two analytes in a sample. For these and other similar questions, separations scientists are turning more often to chemometric tools as a means of visualizing and interpreting the rich data that they obtain from their separations systems.

Here we present a brief overview of separations approaches, with a focus on the data that are derived from different methods and on phenomena in the separations approach that lead to challenges in data interpretation. This is followed by a discussion of approaches that exist for the chemometric interpretation of separations data, specific challenges that arise in the chemometric treatment of these data, and solutions that have been implemented to deal with these challenges.

1.1 Separations Techniques

Chromatography is widely used for the separation, purification, and analysis of mixtures. In general, analytes contained in either a gaseous or liquid mobile phase are flowed past a stationary phase which is usually confined within a column. Depending on the chemistries of the analytes and the conditions of the separation (mobile/stationary phase compositions, temperature, etc.) different compounds will partition between the two phases to varying degrees. The separation arises due to this differential partitioning, with analytes which associate weakly with the stationary phase passing through the column more quickly than those with a greater affinity for the stationary phase (Miller, 2005; Cazes, 2010).

There are many types of chromatography, with the most common being liquid chromatography (LC) where analytes partition between a mobile liquid phase and an immobile stationary phase, and gas chromatography (GC) where the mobile phase is a gas and the stationary phase is a solid or more often a viscous, liquid-like polymer. There are numerous modes for LC separations, including for example reverse-phase (RPLC), normalphase (NPLC), ion (IC), size exclusion (SEC), and hydrophilic interaction (HILIC) to name a few. From a point of view of chemometric data interpretation and the discussion in this chapter, all of these LC separations generate data which are equivalent. In any chromatographic separation, the sample is delivered to the inlet of the column while the outlet is connected to a detector, which records a continuous signal. The detector response rises and then falls to baseline based on the analyte flux passing through it, ideally generating one separate peak with an approximately Gaussian shape for each individual analyte. Assuming that the conditions for repeat analyses are not changed, the peak for a given analyte will appear at the same time in every analysis, with the peak area/height being proportional to the quantity of analyte present in a sample (Poole, 2003; Miller, 2005).

Another separations technique which is popular for some samples is capillary electrophoresis (CE). Here, an electric field applied across a fused silica capillary containing a buffer induces motion of the buffer and analytes in the sample. The CE separation is dependent on differential mobilities of analytes in the solution in the presence of the electric field. This difference in mobilities is based on the fact that different analytes have different charges and sizes in solution. While the separation mechanism of CE is fundamentally different from the chromatographic mechanism, the data are a series of peaks recorded as a function of time. Consequently, the same tools can be applied to data from a CE separation, and similar concerns exist for the interpretation of these data (Poole, 2003; Miller, 2005). For ease of readability, and because chemometrics are more often applied to chromatographic data than electrophoretic data, we will often refer to a chromatogram in this chapter. This could equally be an electropherogram; when considering the application of chemometric techniques to separations data whether the origin is electrophoretic or chromatographic is largely irrelevant.

When tasked with incredibly complex samples, analysts are now turning more and more frequently to so-called comprehensive multidimensional separations (e.g.: GC×GC, LC×LC, CE×CE) (Liu & Phillips, 1991; Erni & Frei, 1978; Michels et al., 2002). In these techniques, the mixture of compounds is sequentially separated by two different separation mechanisms. In the case of GC×GC, for example, a sample might be separated first on an apolar column, followed by a polar column. The exact workings of comprehensive multidimensional separations are beyond the scope of this work, and are

1. INTRODUCTION

discussed elsewhere (Górecki et al., 2004; Cortes et al., 2009; François et al., 2009; Kivilompolo et al., 2011; Li et al., 2011). However, these techniques are gaining in popularity, and are capable of separating exceedingly complex mixtures comprising thousands of individual compounds. Due to the vastly improved separation power of these techniques, the data are much more informationrich, and without some form of chemometric treatment it is essentially impossible to do more than scratch the surface of the information contained therein.

1.2 Separations Data

The detector signal from a separations experiment, when plotted vs. time, yields a series of (ideally) Gaussian peaks, each representing one compound in the sample. Acquisition speed is one consideration for a chromatographic detector: it must be sufficient to faithfully record the profile of each compound as it passes through the detector. In order to obtain an accurate peak profile, the minimum number of acquisition points required across a peak is 10. Thus, the required speed of the detector is intrinsically linked to the nature of the separation. In separations where the base width of the peaks are on the order of 5 s, a data rate of 2 Hz would be acceptable, but when peak widths are 100-200 ms, as in GC×GC, then detector rates on the order of 50-100 Hz are required for quantitative analysis.

From a point of view of chemometric analysis of separations data, another important consideration is whether the detector is univariate or multivariate. Univariate detectors, such as the flame ionisation detector, or single-wavelength UV-visible spectrometer, record only one variable as a function of time, generating data which take the form of a vector of instrument response. Other detectors, typically mass spectrometers and multi-channel spectroscopic instruments, can be operated such that they record a multivariate response. Data from these instruments comprise an array of signal responses with each row representing a time when a response was recorded, and each column representing a variable that was recorded (e.g.: detector wavelength, ion mass-to-charge ratio). To the chemometrician, it is immediately obvious that there are numerous advantages to collecting multivariate chromatographic data; however, it is worth noting that most of this advantage has been by and large ignored by chromatographers. Typically, only the profile of a single variable vs. time would be used to selectively quantify an analyte, or the detector response across all channels at a given time used to help identify a peak.

One other aspect of raw separations data is the sheer number of variables measured for each sample. When a univariate detector is used for a 15 min separation, operating with an acquisition speed of 10 Hz, the data will be a vector of 9000 individual measurements per sample. If a multivariate detector is employed instead, for example a mass spectrometer operating over a 30-300 m/z mass range, this number increases to 2 439 000 individual variables arranged in a 9000 × 271 array per sample! In the case of GC×GC-MS analyses, which are typically 60 min in length but have a high-speed MS collecting data at rates of ~100 Hz, there are on the order of 100 million data points collected for each sample

2. CHALLENGES WITH CHROMATOGRAPHIC DATA

Variations in analytical separations data are, in principle, no different from those derived from any other instrument; being based on both chemical and non-chemical aspects of the analysis. All relevant information will be contained within the chemical variations and any chemometric approach to interpreting chromatographic data must be capable of identifying relevant chemical variation while minimizing the effects of irrelevant chemical and nonchemical variations. Sources of irrelevant chemical variation include matrix peaks, here defined as any chemical source of signal introduced with the sample, but having no bearing on the conclusions drawn from the data. Additionally, there is background signal which can for example derive from changes in mobile phase concentration which influence detector signals in LC or chemical "bleed" signatures from stationary phases as they degrade in GC. Non-chemical variations include, for example, baseline drift (for non-chemical reasons), retention time shifts (due to minor fluctuations in operating conditions), and electronic noise. These may easily interfere with the relevant chemical information, degrading model performance and the validity of results (de la Mata-Espinosa et al., 2011a). Figure 1 presents an overlay of several LC chromatograms of similar samples exemplifying the challenges of baseline drift and retention time shifts. One of the major challenges in handling chromatographic data using chemometric tools is appropriate pre-processing to remove as many non-chemical and irrelevant chemical variations as possible from the data set.

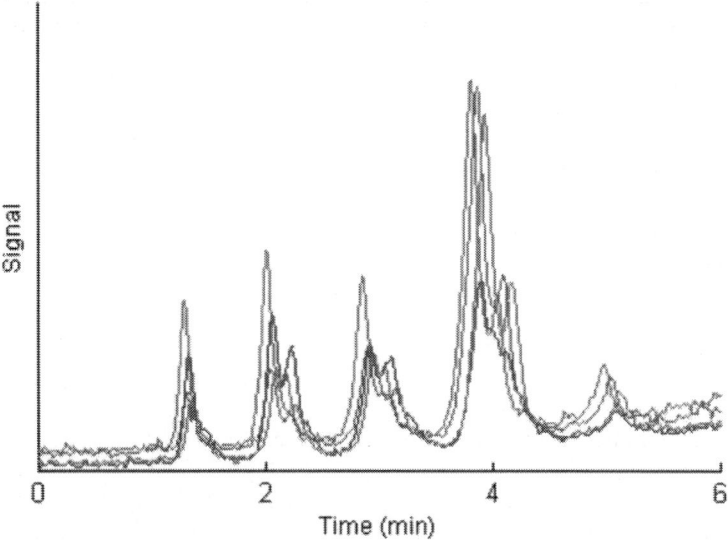

Figure 1. LC chromatograms of edible oils showing a high degree of variation in baseline.

2. CHALLENGES WITH CHROMATOGRAPHIC DATA

Initial efforts into the application of statistical and chemometric tools to chromatographic data were accomplished using data that were processed to provide a list of detected, integrated peak areas or heights (or the calibrated concentrations for known compounds). However, the trend in recent years has turned towards the direct chemometric interpretation of raw chromatographic signals (Watson et al., 2006; Johnson & Synovec, 2002). The reason for this trend is that many errors can occur during integration of raw signals (Asher et al., 2009; de la Mata-Espinosa et al., 2011b). By applying chemometric tools directly to the raw data, many of these errors can be avoided. Of course, when working with the raw data, other issues become more important, most notably retention time shifts and the population of available variables.

2.1 Baseline And Noise

Baseline variations, such as noise and drift, are due to small changes in experimental conditions, for example changes in detector response due to the mobile phase gradient in LC separations or increased levels of stationary phase bleed at higher temperatures in temperature-programmed GC. Other sources of noise and drift could include changes in detector response as its components age, contamination of solvents or gases, and of course electronic noise (which is minimal in modern chromatographic systems).

Chemometric approaches to handling chromatographic data should incorporate baseline correction of some form. When raw chromatographic data are processed, the method of baseline correction and its importance are generally obvious to the analyst. In the case where integrated peak tables are used, this is often done automatically by the chromatographic software with little consideration by the analyst, even though the manner in which the baseline is calculated will significantly influence the determination of peak areas/heights.

2.2 Retention Time Shifts

In all separations, retention times of peaks can easily shift by a few seconds from one analysis to the next. This is not much of an issue with simple samples having only a few peaks which are then integrated prior to chemometric analysis. However, retention times of peaks are used for identifying the compounds. With complex separations, unstable retention times may result in unreliable peak identification, making comparisons from one run to the next impossible. When comparing raw data this is even more important as one must ensure that the peak for a given component is always registered in the exact same position in the data matrix so that the algorithms will recognize the signals correctly.

The causes of retention time shifts depend on the separations technique being used. In GC, peaks may shift due to degradation of the stationary phase, decreasing retention times over time; build-up of heavy matrix components which foul the column, effectively changing the chemistry of the stationary phase; minor gas leaks which alter the flow rate; or even matrix effects on the evaporation rate in the injector, affecting the rate of mass transfer to the column. In LC, peak shifts may be due to small fluctuations in mobile phase chemistry from one run to the next; temperature fluctuations which in turn affect solvent viscosity and solute diffusion coefficients, altering the kinetics as well as the thermodynamics of the separation; or degradation / fouling of the stationary phase of the column. CE is the technique most prone to drastic shifts in migration time, due to the instability of the electroosmotic flow in the capillary (Figure 2). Electroosmotic flow depends on the applied voltage, the buffer concentration and composition, and is incredibly sensitive to the surface chemistry of the capillary. The act of analyzing a sample by CE will often have a minor, possibly irreversible effect on the capillary surface, resulting in a change in the migration time of an analyte.

Shifts in retention times are minimized by proper instrument maintenance, precise control of instrumental conditions or by using approaches such as retention time locking in GC to account for variations in instrument performance (Etxebarria et al., 2009; Mommers et al., 2011) and relative retention times in CE. Even with these approaches, some retention time shifting will occur and require more advanced alignment techniques for correction prior to chemometric analysis.

2.3 Incomplete Separation

Another challenge with the interpretation of chromatographic data is incomplete separation of peaks. If two or more compounds have similar retention characteristics under a given set of separation conditions, they will not be completely resolved, as evidenced by the peak clusters in Figure 1. In these cases, apportioning the signal between the different compounds becomes a challenge, especially for univariate signals. The general approach used for these cases is one of deconvolution: decomposing the analytical signal to determine the contribution of each coeluting compound, or to determine the contribution of the compound of interest, disregarding the remaining data.

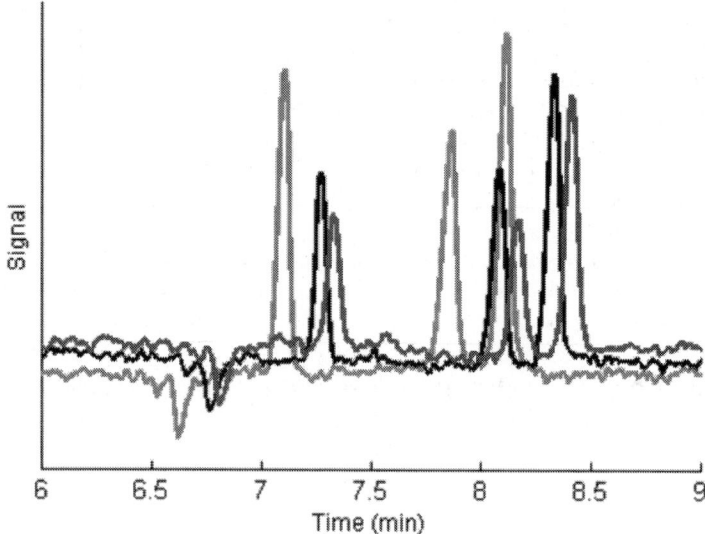

Figure 2. CE of substituted benzenes showing extreme misalignment.

2.4 Data Overload

As shown in Section 1.2, raw chromatographic signals present an overabundance of data to the analyst. This poses several challenges. From a practical point of view, attempts to construct a chemometric model using the entirety of the data set could easily exceed the capabilities of the computer system being used. More fundamentally, if the raw data are considered, the number of variables measured for each sample will vastly outnumber the number of samples available in the data set. These overdetermined systems can defeat many chemometric techniques due, for example, to collinear variables. Finally, for most chromatograms, especially multidimensional ones, only a small fraction of the data points actually contain meaningful signal. Most of the signal is due to background noise or irrelevant matrix components. Consequently, the raw data must somehow be reduced in size prior to chemometric analysis. This is typically achieved via a feature selection approach, as discussed in Section 3.3.3.

3. PRE-PROCESSING STEPS FOR CHROMATOGRAPHIC DATA

3.1 Baseline Correction

The aim of baseline correction is to separate the analyte signal of interest from signal which arises due to changes in mobile phase composition or stationary phase bleed and signal due to electronic noise. Several baseline correction methods have been proposed in literature, with the two most common approaches being to fit a curve to the data and subtract this value from the signal, and modeling the baseline to exclude it using factor models (Amigo et al., 2010).

Curve fitting is the classical approach used in virtually all commercial software packages provided by vendors of separations equipment. The algorithms used in this approach fit a polynomial function across segments of the chromatogram using regions where no analyte peaks elute to determine the coefficients of the polynomial and then interpolating the background signal for regions where peaks are eluting. The functions are usually first-order polynomials; however, higher-order polynomials or a series of connected first-order polynomials are also used in some situations. Having determined the equation of the background signal, the fitted line is then subtracted from the signal (Brereton, 2003; Gan et al., 2006; Kaczmarek et al., 2005; Zhang et al., 2010; Persson & Strang, 2003; Eilers, 2003). Correction of the baseline using curve fitting is demonstrated in Figure 3.

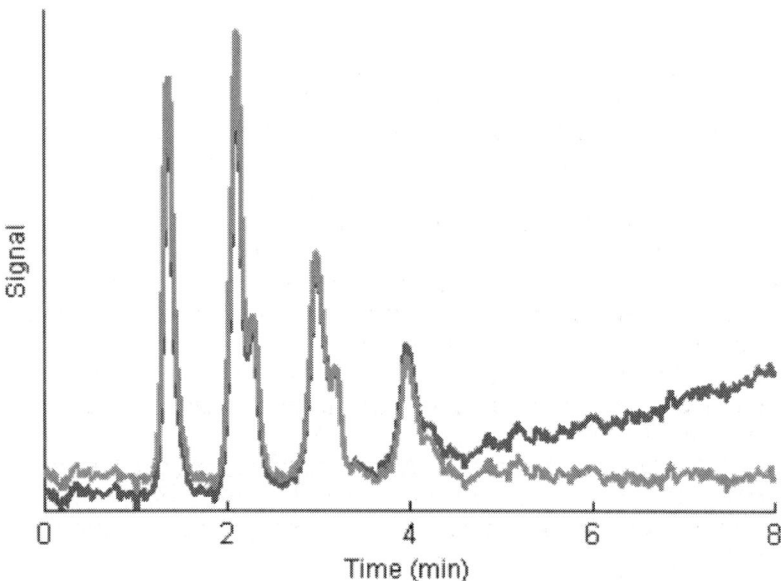

Figure 3. An LC chromatogram before (blue) and after (red) baseline correction.

The approach of using models such as parallel factor analysis (PARAFAC) for background correction is analogous to the use of these approaches for deconvoluting coeluting peaks. As these models are more often used for this purpose than for simple background correction, they will be discussed in more detail in Section 3.3. These approaches often rely on having a multivariate signal and are applied to the chromatogram or more typically small selected regions where a single analyte elutes. The result of applying these deconvolution techniques for background correction is essentially the deconvolution of a single analyte peak, with the background noise making up the error matrix (Amigo et al., 2010). These approaches are generally more powerful and likely result in better quality analytical data, but they are not widely used in separation science. The reason for this is likely historical as these tools have only recently become available to the separation sciences, while the classical curve fitting approach is well established, works with univariate detectors, and performs well in most practical situations.

3.2 Alignment of Separations Data

The retention times of analytes in separations fluctuate from one analytical run to the next and, in order for chemometric techniques to be applied to separations data, these fluctuations must be corrected during pre-processing. This ensures that the signal from each analyte in each analysis is correctly registered within the data matrix to be processed. There are essentially two approaches to this problem: integrated peak tables, or mathematical warping and alignment of the raw signal.

3. PRE-PROCESSING STEPS FOR CHROMATOGRAPHIC DATA

3.2.1 Peak tables

Integrated peak tables are the simplest way to ensure that analytical separations data are properly aligned for chemometric processing. In order to use this approach, one must be able to reliably assign a unique identifier to each peak in each sample of the data set, and ensure that the same compound is identified with the same identifier in each sample. It should be noted that while the compound name is an obvious identifier, a series of labels such as Unknown x, where x is a numerical identifier would also be acceptable in the event that compound names were unknown, so long as compounds are matched correctly. Rather than identifying peaks by retention time, one could use relative retention times or retention indices in order to adjust for slight variations in the retention times of peaks. Algorithms for aligning peak tables exist and perform well, so long as some peaks can be easily and reliably matched across all chromatograms (Lavine et al., 2001).

The challenges with this approach stem from its reliance on integrated peak tables. Thus, any integration errors due to poorly-resolved peaks or peaks that are missed due to falling outside of integration parameters in the software will impact any subsequent analysis.

3.2.2 Raw signal alignment

Alignment of raw chromatographic signals prior to chemometric processing is more complex than the alignment of peak tables. In addition to the three more popular algorithms that will be presented below, there are several others that have been developed (Yao et al., 2007; Toppo et al., 2008; Eilers, 2004; Van Nederkassel et al., 2006). In deciding which approach to use, one of the first questions to be answered is if the analysis is to be qualitative or quantitative. This is because some alignment methods can distort peaks, affecting their quantification. Some of the more common algorithms include correlation optimized warping (COW) (Nielsen et al., 1998; Tomasi et al., 2004), correlation optimized shifting (coshift) (Van den Berg, 2005), and a piecewise peak-matching algorithm (Johnson et al., 2003).

In instances where there are non-systematic peak shifts, COW is a popular algorithm. COW relies on stretching or compressing segments of a sample signal such that the correlation coefficient between it and a reference signal is maximized for each interval. Care must be taken with the selection of the input parameters to avoid significant changes in peak shapes as this approach to the warping of the chromatogram has been shown to affect peak areas, leading to poor quantitative conclusions (Nielsen et al., 1998; Tomasi et al., 2004).

A fast and simple alignment algorithm is coshift. This algorithm is useful when data only require a single left-right shift in retention time. The entire data matrix is shifted in one direction or the other by a set amount, maximizing the correlation between a target and the data matrix that required alignment. The single shifting value for the entire data matrix is a weakness, especially for chromatographic data where peaks can shift in different directions and to different extents in a single file. To handle this, an algorithm termed icoshift (interval-correlation-shifting) has been derived from coshift. Icoshift aligns each data matrix to a target by maximizing the cross-correlation between the sample and the target within a series of user-defined intervals (Savorani et al., 2010). The use of multiple intervals permits the alignment of separations data where shifts of different magnitudes and directions occur. These alignment algorithms have been used successfully for both one-dimensional data (de la Mata-Espinosa, 2011a; Liang, 2010; Laursen, 2010) and two-dimensional data, with some modifications (Zhang, 2008). It is important to note that the shifting of chromatograms using coshift or icoshift does not lead to distortions of peak shape, and consequently does not introduce errors into quantitative results.

The piecewise peak matching approach (Johnson et al., 2003) provides another avenue for chromatographic alignment. In this approach, peaks are identified in a target signal to which all other signals will be aligned. The algorithm then identifies peaks within the sample signals located within predetermined windows of the peaks in the target. Peaks within windows are deemed to come from the same compound, and matched. The chromatograms are aligned by stretching or compressing the regions between peak apexes. A variant of this algorithm can be used when MS data are available. In this case, the mass spectrum at the apex of each peak in the target signal is compared to the mass spectrum of each peak within a set window on the sample signal and peaks are matched if their spectra have a high enough match quality (Watson et al., 2006). A general scheme for peak alignment using this approach is described in Figure 4. Depending on the number and relative positions of the peaks in chromatograms matched using this approach, peak shapes may be altered, possibly affecting quantitative results.

One of the biggest challenges for all alignment algorithms is that they depend on the data to be aligned being reasonably similar in terms of both matrix and analyte peaks. In some instances this will not be the case. In our laboratory, we have observed this when analyzing arson debris where the matrix and analytes form an incredibly complex and variable chromatogram from one sample to the next. A similar situation can be easily imagined when processing samples of biological origin. One solution to this issue is to add markers to every sample prior to the separation step in the analysis. These markers should be easily identifiable within the samples, even under conditions where they coelute with matrix components; should occur in multiple, evenly distributed locations along the chromatogram, and should not occur natively in the samples. One choice is a series of deuterated compounds which, with MS detection, are trivial to identify even in a complex mixture (Sinkov et al., 2011b). One additional benefit is that these compounds can act as internal standards if quantitative results are desired.

3. PRE-PROCESSING STEPS FOR CHROMATOGRAPHIC DATA

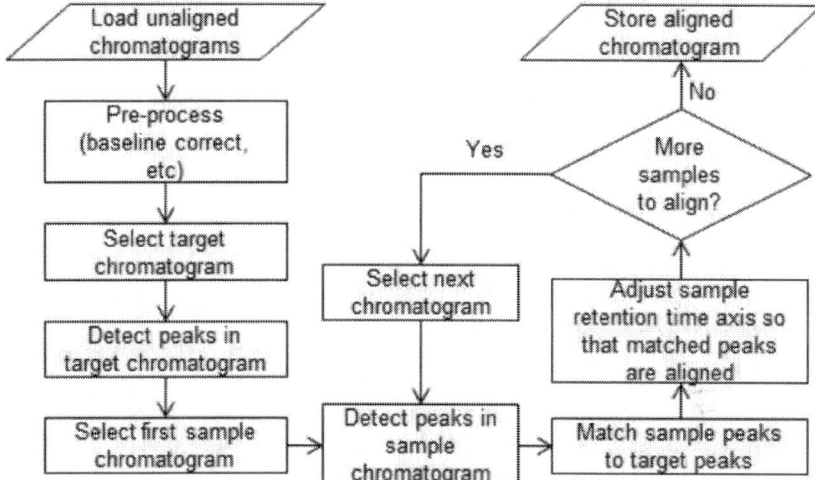

Figure 4. Flowchart for target-based chromatographic alignment, adapted from (Johnson et al., 2003).

3.3 Deconvolution of Overlapping Peaks

The central issue in deconvolution is depicted in Figure 5. The instrument response is represented as a black solid line which is the sum of the four dashed, coloured peaks. Ideally, the four signals should be individually quantified. This is a common problem for analytical separations, even those of relatively simple mixtures. Some of these issues may be solved by changing the experimental conditions or using characteristic features (wavelengths or ions) of the coeluting analytes and a multivariate detector to selectively detect and quantify them. However, in many cases this is insufficient and more advanced techniques must be used. The strategies used for deconvolution depend heavily on whether the detector signal is univariate or multivariate.

3.3.1 Deconvolution of univariate signals

In the case of univariate signals, one is typically limited to using univariate curve-fitting analyses where a number of Gaussian or modified Gaussian curves are determined such that the sum of these curves fits the experimentally observed cluster of peaks (Felinger, 1994). In these approaches, only a small window of chromatographic data (one peak cluster) should be processed at a time, and constraints such as fixed peak widths, shapes, unimodality, and non-negativity are often required to ensure the validity of the solution.

To solve a univariate deconvolution problem, approaches such as evolving factor analysis (EFA) (Maeder, 1987) or multivariate curve resolution (MCR) (Tauler & Barceló, 1993), among others (Vivó-Truyols et al., 2002; Sarkar et al., 1998; Kong et al. 2005) can be used. When these approaches are used with univariate data, the variables to be solved for are the number, positions, and abundances of each of the peaks that make up the signal.

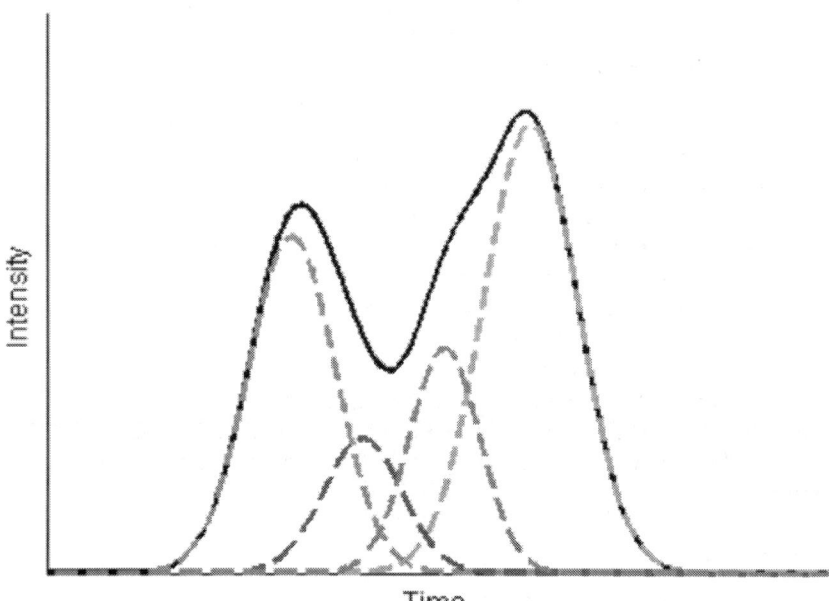
Figure 5. Deconvolution of overlapping peaks. The black, solid trace represents the analytical signal observed at the detector, which is the sum of the four peaks represented by dashed lines.

Multivariate curve resolution is widely applicable to separations data and is one of the most common approaches (Franch-Lage et al., 2011; Marini et al., 2011, de la Mata-Espinosa et al., 2011a). The aim of this technique is to determine the number of components present in a sample and the contribution of each component to the sample. In performing MCR, the concentration and response profiles for each analyte are obtained, providing a qualitative and semi-quantitative overview of the components in an unresolved mixture without a priori knowledge of the mixture composition.

3.3.2 Deconvolution of multivariate signals

When multivariate detectors are used for separations, the additional dimension of information can be exploited to aid in deconvolution. MCR and EFA can also be used with multivariate data. In the case of MCR, the experimental matrix is decomposed into a matrix of concentration vs. time profiles (deconvoluted peaks) and pure spectral profiles of each compound. Knowledge of the number of components contributing to the signal in the region being deconvoluted is useful to guide the process and improve the results (de Juan & Tauler, 2006), though strictly speaking it is not required.

Parallel factor analysis (PARAFAC) (Harshman, 1970; Bro, 1997; Amigo et al., 2010) is a technique that is ideally suited for interpreting multivariate separations data. PARAFAC is a decomposition model for multivariate data which provides three matrices, A, B and C which contain the scores and loadings for each component. The residuals, E, and the number of factors, r, are also extracted. The PARAFAC decomposition finds the best trilinear model that minimizes the sum squares of the residuals in the model through a procedure of alternating least squares.

The biggest advantage of using PARAFAC over other models is the uniqueness of the solution; PARAFAC is less flexible and uses fewer degrees of freedom, being a more restricted model. However, its unique solution reflects actual pure analyte profiles in both the time dimension and the spectral dimension. Thus, the results of PARAFAC analysis on a cluster of overlapping multivariate peaks provide both qualitative and quantitative data where the deconvoluted signals appear as analyte peaks. One restriction to the use of PARAFAC is that the data must be trilinear (Bro, 1997; Amigo et al., 2010). In the case of chromatographic techniques with a multivariate detector, the dimensions are retention time, detector signal, and samples. In the case of comprehensive multidimensional separations, such as GC×GC, PARAFAC considers retention in the two dimensions and the samples as the three dimensions.

3.4 Feature Selection

High data acquisition rates combined with the length of time required for many separations results in a large number of data points collected for a given separation (see Section 1.2). In many situations, most of the data are collected when no analytes are eluting from the system, and represent background signal when only mobile phase is reaching the detector. In the case of spectroscopic and especially mass spectral detectors, at a given point in time, many of the recorded data in this dimension will not contain useful information, even when an analyte of interest is eluting. Furthermore, many components in the mixture can be completely irrelevant to analysis (Johnson & Synovec, 2002; Sinkov & Harynuk, 2011a). Consequently, only a small portion of separations data is potentially useful. It is also well known that any model will be heavily influenced by the specific variables that are included in its construction (Kjeldahl & Bro, 2010).

The inclusion of irrelevant data is detrimental to the model because the mathematics attempt to account for variations observed in these irrelevant variables. Consequently the model is forced to model noise, resulting in a decrease in its predictive ability. Worse yet, the model could fit the data well and provide a seemingly useful prediction, until crossvalidation shows otherwise. Finally, the inclusion of extraneous variables increases the demands on the computer system being employed, making model construction slower, or in some cases outright impossible. Thus, prior reduction of separations data to a manageable size is crucial. Figure 6 depicts situations where either too few or too many variables were used to model a system.

One common manner to achieve data reduction is to use a table of integrated peaks instead of raw chromatographic data. This has the advantage of reducing the number of variables to those compounds included in the peak list, removing baseline noise and, if the analyst knows which exact peaks to use, removing signal from irrelevant compounds. Problems with this approach include the restriction to identified compounds, which may or may not include all of the information required for modeling, and integration errors that skew results. Finally, even with an error-free comprehensive peak table, the analyst must still perform feature selection since many peaks will undoubtedly be irrelevant to the analysis.

In the case of multivariate detection, it can be advantageous to monitor only one or a few channels (wavelengths, ions, etc.) as this will selectively detect only a portion of the analytes, allowing the analyst to avoid many interfering species while greatly reducing the size of the data. However, in these cases the analyst must know exactly what signals to use and runs the risk of missing important features of the data encoded in the channels that were ignored. Further, using this approach destroys much of the multivariate advantage that can be realized through using these more complex (and expensive) detection strategies.

Objective feature selection techniques generally have two steps: variable ranking, and variable selection. Objective variable ranking techniques such as analysis of variance (ANOVA) (Johnson & Synovec, 2002), the discriminating variable test (DIVA) (Rajalahti et al., 2009a, 2009b), and informative vectors (Teofilo et al., 2009) have the distinct advantage that variables are ranked based on a mathematically calculable "perceived utility" and not on subjective analyst perception. In essence, the data are given the chance to inform the user of what is relevant and what is likely noise, providing an approach that can be generalized to any set of analytical data.

ANOVA is an effective method when the goal is to discriminate between classes of samples. ANOVA calculates the F ratio for each variable: the ratio of between-class variance to within-class variance. If the F ratio for a given variable is high, it is deemed to be more valuable for describing the difference between classes. Once the F ratio has been calculated for every data point in the chromatogram, the variables can be ranked in order of decreasing F ratio. A chemometric model is then constructed using a fraction of variables having the highest F ratio. One significant advantage of ANOVA is that the algorithm can be written with memory conservation in mind and thus is easily applied to data sets with very large numbers of samples and variables (hundreds or thousands of samples, each containing millions of variables). Consequently, it can be easily applied to a set of GC-MS chromatograms across the entire chromatogram, something that is difficult for other feature ranking approaches.

DIVA is a feature ranking technique that aids feature selection prior to chemometric analysis (Rajalahti et al., 2009a, 2009b). This approach involves the creation of a PLS-DA model using all candidate variables. Projecting this PLS-DA model onto a new single LV yields what is termed a target projected (TP) model (Rajalahti et al., 2009a). From this, the ratio of explained variance to residual variance for each variable in the TP model provides its selectivity ratio, upon which variables are ranked (Rajalahti et al., 2009a, 2009b; Kvalheim, 1990; Kvalheim & Karstang, 1989). DIVA produces a ranking that is slightly different than that produced by ANOVA, though a direct comparison on chromatographic data has not yet been performed to our knowledge.

Once variables have been ranked, those to be included in the model must be selected. This is generally achieved by constructing a model using a forward-selection or backwards elimination approach, in an attempt to maximize some metric of model quality. Model quality can be assessed based on several metrics such as mean correct classification rates (Rajalahti et al., 2009b) or the degree of separation between classes of samples in principal component (or latent variable) space, for example using either a Euclidian distance-based metric (Pierce et al., 2005) or a metric that accounts for size and shape of clusters (Sinkov & Harynuk, 2011a).

The one exception to the rank-and-select approach are genetic algorithms (Yoshida et al., 2001), though due to the sheer number of variables present in a typical separation, these are not often used on the raw separations data as arriving at the optimal number and combination of variables is computationally inefficient and uncertain.

Sometimes, several feature selection methods are used for a given analysis. For example, an analyst might reduce chromatogram to a peak table, selecting a series of candidate variables of interest and then perform further variable ranking and optimization on the integrated peak table, especially in the case of multidimensional separations where hundreds, if not thousands of compounds can be resolved (Felkel et al., 2010).

Finally, cross-validation is extremely important, especially when processing raw separations data and using a feature ranking approach such as ANOVA. As discussed previously, raw separations data contain on the order of 105 to 106 data points for each sample. In these cases of overdetermined systems it is entirely possible that some combinations of variables containing only noise will, by random chance, indicate a difference between samples. When handling raw separations data, a good approach to avoid this problem is to break the data set into three separate sets: a training set to construct the model, an optimization set to optimize data processing parameters (such as alignment and feature selection), and finally a test set to determine if the optimized model has any meaning (Brereton, 2007). Of course this does require that one collect data for a large number of samples so that a representative population of samples exists for each of the three subsets of data.

4. APPLICATIONS AND EXAMPLES

After applying the appropriate pre-processing, different chemometric techniques can be applied according to the aim of the study. Pattern recognition is one of the chemometric methods most used in analytical chemistry and this is true for separations data. Pattern recognition can be generally divided into two classes: exploratory data analysis and unsupervised and supervised pattern recognition (Otto, 2007; Brereton, 2007).

Exploratory data analysis aims to extract important information, detect outliers and identify relationships between samples and its use is recommended prior to the application of other chemometric techniques. Examples of the use of exploratory data analysis tools applied to separations data include principal component analyisis (PCA) (de la Mata-Espinosa et al., 2011a; Ruiz-Samblas et al., 2011) and factor analysis (Stanimirova et al., 2011).

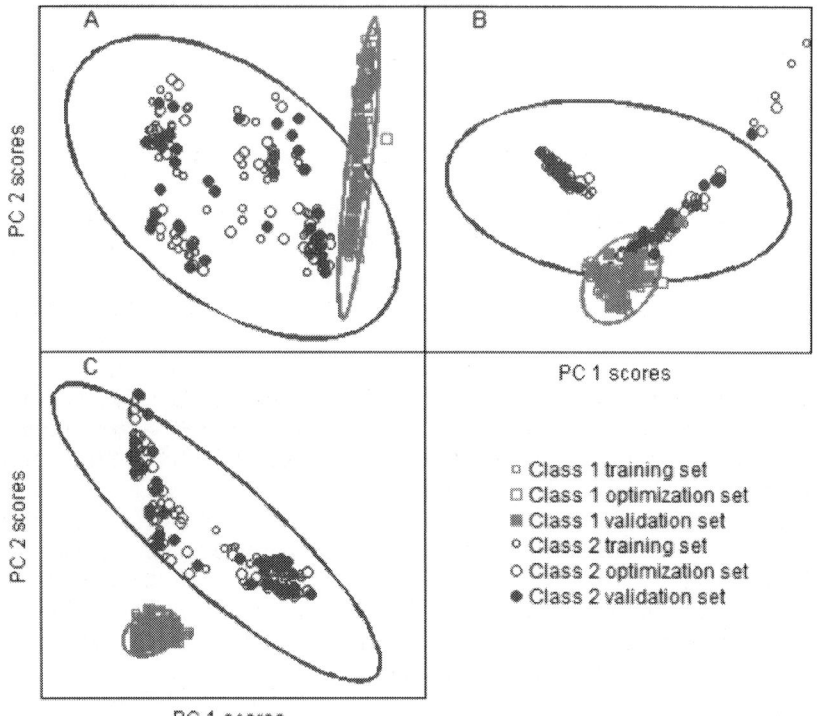

Figure 6. Models constructed from the same data set using different numbers of top-ranked variables. (A) Too few variables; (B) Too many variables; (C) Optimal number of variables.

Unsupervised pattern recognition techniques uncover patterns within a data set without a priori class assignment of samples. Here, the objective is to find patterns in the data which allow grouping of similar samples using, for example, cluster analysis which has been applied to separations data by Reid et al. (2007). When supervised pattern recognition is used, the classes of samples in a training set are known and used to calibrate a model, which is then used to predict class assignments of unknown samples. Some examples of which are linear discriminant analysis (LDA), and partial least squares-discriminant analysis (PLS-DA) (de la Mata-Espinosa et al., 2011b; Zorzetti et al., 2011; Sinkov et al., 2011b). In a study performed by Sinkov et al., two alignment techniques for chromatographic data were compared. The data comprised raw GC-MS chromatograms of simulated arson debris where some samples contained different types of gasoline weathered to different extents spiked into debris samples which themselves exhibited a high degree of variability in their chemical composition. The goal was to build a PLS-DA model that could correctly classify debris samples based on whether or not they contained gasoline (Figure 7). As can be seen, the alignment algorithm used has a direct

4. APPLICATIONS AND EXAMPLES

impact on the quality of the predictions. In Figure 7A, there are multiple false positives, false negatives, and ambiguous samples. In Figure 7B, all samples are classified correctly and there are no ambiguous samples.

Another example of applying chemometrics to separations data is depicted in Figures 8 and 9. Here, interval PLS (iPLS) was applied to blends of oils in order to quantify the relative concentration of olive oil in the samples (de la Mata-Espinosa et al., 2011b). iPLS divides the data into a number of intervals and then calculates a PLS model for each interval. In this example, the two peak segments which presented the lower root mean square error of cross validation (RMSECV) were used for building the final PLS model.

As mentioned in Section 3.3.2, PARAFAC is a chemometric tool for multidimensional data treatment. The scores and loadings obtained with PARAFAC can be used in two-way models for data exploration and quantitative analysis (Vosough et al., 2010). When small deviations in trilinearity exist within the data, usually due to relatively small shifts in retention time in the case of separations data, a modified version of PARAFAC called PARAFAC2 is recommended for use (Bro et al., 1999).

Like PARAFAC, PARAFAC2 decomposes raw data into loading and score matrices but without the imposition of trilinearity as in PARAFAC. Even without this constraint, the PARAFAC2 model preserves the property of uniqueness that is so advantageous with PARAFAC. Thus, analyte profiles and concentrations can be estimated by PARAFAC2 even if chromatographic alignment is not perfect (Amigo et al., 2008; Skov et al., 2009).

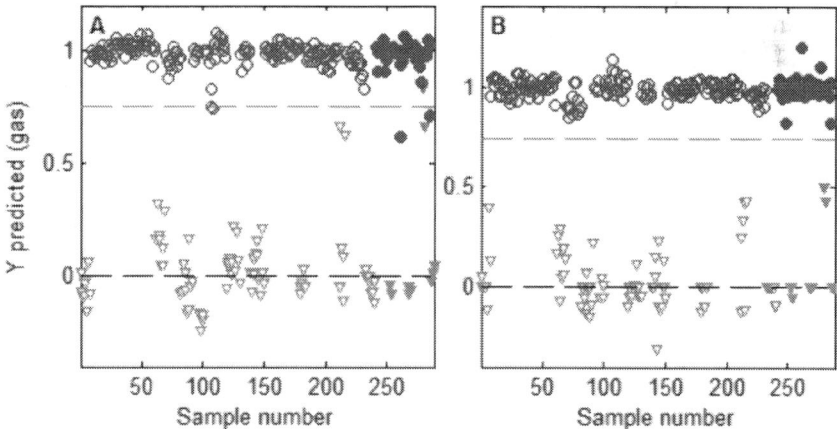

Figure 7. PLS-DA Models for identifying gasoline in simulated arson debris derived from the same raw data, but aligned with different techniques. (A) Feature-based alignment; (B) Deuterated alkane ladder – based alignment. All other treatment and model construction algorithms were the same in both cases. Hollow markers indicate data in the training set while filled markers indicate data in the validation set. Circles represent debris containing gasoline while triangles represent gasoline-free debris. Reprinted from Sinkov et al., 2011b, with permission.

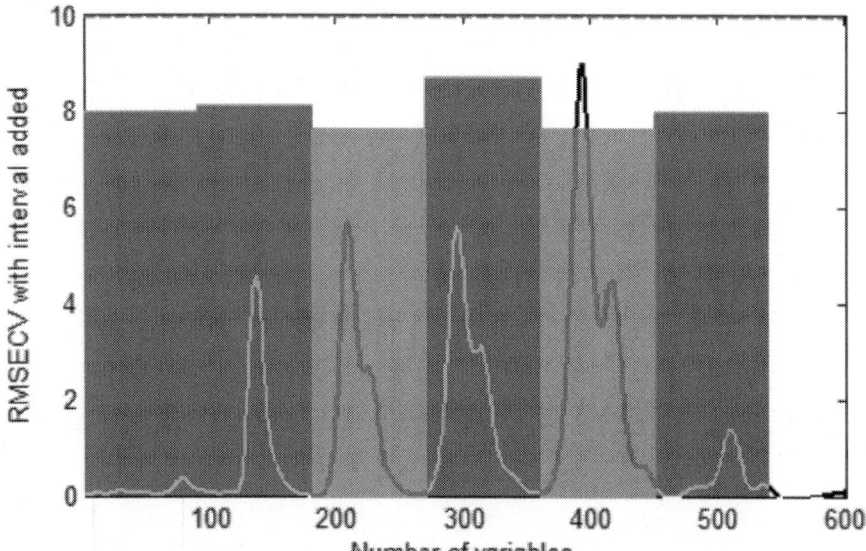

Figure 8. Feature selection using iPLS. Segments in green showed lower RMSECV and were thus used to construct the final model. Reprinted from de la Mata-Espinosa et al., 2011b, with permission.

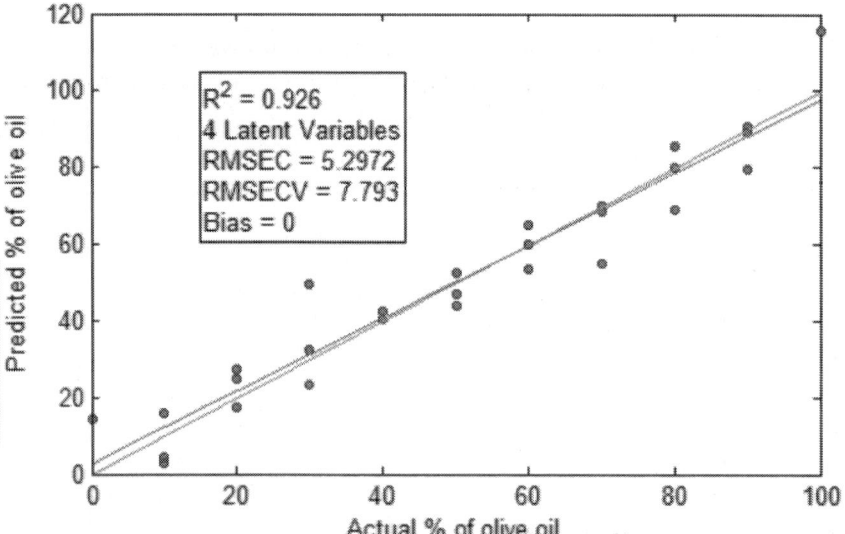

Figure 9. Predicted vs. actual % olive oil using PLS model constructed based on results in Figure 8. Reprinted from de la Mata-Espinosa et al., 2011b, with permission.

5. CONCLUSIONS

The analyst must choose from a plethora of methods for processing separations data, a potentially daunting task. It is our hope that this review will help chromatographers entertaining thoughts of applying chemometrics to their data understand what they must consider when choosing how to prepare their data. Likewise, it is hoped that we have informed chemometricians of some of the specific challenges associated with the processing of chromatographic data and the origins of those limitations. In the development of a chemometric model for the interpretation of separations data, there are numerous opportunities for missteps that will exclude key information from the model and/or generate meaningless results. However, when due care is taken there are also many opportunities to apply chemometric techniques to transform the rich data generated by these powerful analytical tools into valuable information effectively and efficiently.

REFERENCES

1. Amigo, J.M.; Skov, T.; Bro, R.; Coello, J. & Maspoch, S. (2008). Solving GC-MS problems with PARAFAC2. Trends in Analytical Chemistry, Vol.27, No.8, (September 2008), pp. 714- 725, ISSN 0165-9936
2. Amigo, J.M.; Skov, T. & Bro, R. (2010). ChroMATHography: solving chromatographic issues with mathematical models and intuitive graphics. Chemical Reviews, Vol.110, No.8, (May 2010), pp. 4582-4605, ISSN 1520-6890
3. Asher, B.J.; D'Angostino, L.A.; Way, J.D.; Wong, C.S. & Harynuk, J.J. (2009). Comparison of peak integration methods for the determination of enantiomeric fraction in environmental samples. Chemosphere, Vol.75, No.8, (May 2009), pp. 1042-1048, ISSN 0045-6535
4. Brereton, R.G. (2003). Chemometrics Data Analysis for the Laboratory and Chemical Plant, Wiley, ISBN 0-474-78977-8, UK
5. Brereton, R.G. (2007). Applied Chemometrics for Scientists, John Wiley & Sons Inc., ISBN 978-0- 470-01686-2, Toronto, Canada
6. Bro, R. (1997). PARAFAC. Tutorial and applications. Chemometrics Intelligent Laboratory Systems, Vol.38, No.2, (October 1997), pp. 149-171, ISSN 0169-7439
7. Bro, R.; Andersson, C.A.& Kiers, H.A.L. (2009). PARAFAC-Part II. Modeling chromatographic data with retention times shifts. Journal of Chemometrics, Vol.13, No.3-4, (May-August 1999), pp. 295-309, ISSN 0886-9383
8. Casez, J. (2010). Encyclopaedia of Chromatography, (3rd ed.) CRC Press, ISBN 1-4200-8483, Florida, USA

9. Cortes, H.J.; Winniford, B.; Luong, J. & Pursch, M. (2009). Comprehensive two dimensional gas chromatography review. Journal of Separation Science, Vol.32, No.5-6, (March 2009), pp. 883-904, ISSN 1615-9306
10. de Juan, A. & Tauler, R. (2006). Multivariate Curve Resolution (MCR) from 2000: Progress in concepts and applications. Critical Reviews in Analytical Chemistry, Vol.36, No.3-4, (2006) pp. 163-176, ISSN 1040-8347
11. de la Mata-Espinosa, P.; Bosque-Sendra, J.M.; Bro, R. & Cuadros-Rodríguez, L. (2011a). Discriminating olive and non-olive oils using HPLC-CAD and chemometrics. Analytical and Bioanalytcial Chemistry, Vol.399, No.6, (February, 2011), pp. 2083-2092, ISSN 1618-2650
12. de la Mata-Espinosa, P.; Bosque-Sendra, J.M.; Bro, R. & Cuadros-Rodríguez, L. (2011b). Olive oil quantification of edible vegetable oil blends using triacylglycerols chromatographic fingerprints and chemometric tools. Talanta, Vol.85, No.1, (July 2011), pp. 183-196, ISSN 0039-9140
13. Eilers, P.H.C. (2003). A perfect Smoother. Analytical Chemistry, Vol.75, No.14, (July 2003) pp. 3631-3636, ISSN 0003-2700
14. Eilers, P.H.C. (2004). Parametric Time Warping. Analytical Chemistry, Vol.76, No.2, (January 2004), pp. 404-411, ISSN 0003-2700
15. Erni, F. & Frei, R.W. (1978). 2-Dimensional column liquid-chromatographic technique for resolution of complex mixtures. Journal of Chromatography, Vol.149, (February 1978), pp. 561-569 ISSN 0021-9673
16. Etxebarria, N.; Zuloaga, O.; Olivares, M.; Bartolomé. L.J. & Navarro, P. (2009). Retentiontime locked methods in gas chromatography. Journal of Chromatography A, Vol.1216, No.10, (March 2009), pp. 1624-1629 ISSN 0021-9673
17. Felinger, A. (1994). Deconvolution of overlapping skewed peaks. Analytical Chemistry, Vol.66, No.19, (October 1994), pp. 3066-3072, ISSN 0003-2700
18. Felkel, Y.; Dorr, N.; Glatz, F. & Varmuza, K. (2010). Determination of the total acid number (TAN) of used gas engine oils by IR and chemometrics applying a combined strategy for variable selection. Chemometrics and Intelligent Laboratory Systems, Vol. 101, No. 1, (March, 2010), pp. 14-22 ISSN 0169-7439
19. Franch-Lage, F.; Amigo, J.M.; Skibsted, E.; Maspoch, S. & Coello, J. (2011). Fast assessment of the surface distribution of API and excipients in tablets using NIR-hyperspectral imaging. International Journal of Pharmaceutics, Vol.441, No.1-2, (June 2011), pp. 27- 35, ISSN 0378-5173

20. François, I.; Sandra, K. & Sandra, P. (2009). Comprehensive liquid chromatography: Fundamental aspects and practical considerations—A review. Analytica Chimica Acta, Vol.641, No.1-2, (May 2009), pp. 14-31, ISSN 0003-2670
21. Gan, F.; Ruan, G. & Mo, J. (2006). Baseline correction by improved iterative polynomial fitting with automatic threshold. Chemometrics and Intelligent Laboratory Systems, Vol.82, No.1 (May 2006), pp. 59-65, ISSN 0169-7439
22. Górecki, T.; Harynuk, J. & Panić, O. (2004). The evolution of comprehensive twodimensional gas chromatography, Journal of Separation Science, Vol.27 (2004) pp. 359-379, ISSN 1615-9306
23. Harshman, R.A. (1970). Foundations of the PARAFAC procedure: models and conditions for an 'exploratory' multimodal factor analysis. UCLA Working Papers Phonet. Vol 16, (1970), pp. 1-84
24. Johnson, K.J. & Synovec, R.E. (2002). Pattern recognition of jet fuels: comprehensive GC×GC with ANOVA-based feature selection and principal component analysis. Chemometrics and Intelligent Laboratory Systems, Vol.60, No.1-2, (January 2002), pp. 225-237, ISSN 0169-7439
25. Johnson, K.J.; Wright, B.W.; Jarman, K.H. & Synovec, R.E. (2003). High-speed peak matching algorithm for retention time alignment of gas chromatographic data for chemometric analysis. Journal of Chromatography A, Vol.996, No.1-2, (May 2003), pp. 141-155, ISSN 0021-9673
26. Kaczmarek, K.; Walczak, B.; de Jong, S. & Vandeginste, B.G.M. (2005). Baseline reduction in two dimensional gel electrophoresis images. Acta Chromatographica, Vol.15 (2005), pp. 82-96, ISSN 1233-2356
27. Kivilompolo, M.; Pol, J. & Hyotylainen, T. (2011). Comprehensive two-dimensional liquid chormatography (LC×LC): A review. LC GC Europe, Vol.24, No 5 (May 2011), pp. 232-+, ISSN 1471-6577
28. Kjeldahl, K. & Bro, R. (2010). Some common misunderstandings in chemometrics. Journal of Chemometrics, Vol.24, No.7-8, (July-August, 2011), pp. 558-564, ISSN 0886-9383
29. Kong, K.; Ye, F.; Guo, L.; Tian, J. & Xu, G. (2005). Deconvolution of overlapped peaks based on the exponentially modified Gaussian model in comprehensive two-dimensional gas chromatography, Journal of Chromatography A, Vol.1086, No.1-2 (September 2005) pp. 160-164, ISSN 0021-9673
30. Kvalheim, O.M. & Karstang, T.V. (1989). Interpretation of latent-variable regression models. Chemometrics and Intelligent Laboratory Systems, Vol.7, No.1-2, (December 1989), pp. 39-51, ISSN 0169-7439

31. Kvalheim, O.M. (1990). Latent-variable regression models with higher-order terms: An extension of response modelling by orthogonal design and multiple linear regression. Chemometrics and Intelligent Laboratory Systems, Vol.8, No.1, (May 1990), pp. 59-67, ISSN 0169-7439
32. Lavine, B.K.; Brzozowski, D.; Moores, A.J.; Davidson, C.E. & Mayfield, H.T. (2001). Genetic algorithm for fuel spill identification. Analytica Chimica Acta, Vol.437, No.2, (June 2001), pp. 233-246, ISSN 0003-2670
33. Laursen, K.; Frederiksen, S.S.; Leuenhagen, C. & Bro, R. (2010). Chemometric quality contrl of chromatographic purity. Journal of Chromatography A, Vol.1217, No.42 (October 2010), pp. 6503-6510, ISSN 0021-9673
34. Li, Y.H.; Wojcik, R & Dovichi, N.J. (2011). A replaceable microreactor for on-line protein digestion in a two dimensional capillary electrophoresis system with tandem mass spectrometry detection. Journal of Chromatography A, Vol.1218, No.15 (April 2011), pp. 2007-2011, ISSN 0021-9673
35. Liang, Y.; Xie, P. & Chau, F. (2010). Chromatographic fingerprinting and related chemometric techniques for quality control of traditional Chinese medicines. Journal of Separation Science, Vol.33, No.3 (February 2010), pp. 410-421, ISSN 1615- 9314
36. Liu, Z. & Phillips, J.B. (1991). Comprehensive 2-dimensional gas-chromatography using a modulator interface. Journal of Chromatographic Science, Vol.29, No.6 (June 1991), pp. 227-231, ISSN 0021-9665
37. Maeder, M. (1987). Evolving factor analysis for the resolution of overlapping chromatographic peaks. Analytical Chemistry, Vol.59, No.3, (February 1987), pp 527- 530, ISSN 0003-2700
38. Marini, F.; D'Aloise, A.; Bucci, R.; Buiarelli, F.; Magri, A.L. & Magri, D. (2011), Fast analysis of 4 phenolic acids in olive oil by HPLC-DAD and chemometrics, Chemometrics and Intelligent laboratory systems, Vol.106, No.1, (March 2011), pp. 142-149, ISSN 0169- 7439
39. Michels, D.A.; Hu, S.; Schoenherr, R.M.; Eggertson, M.J. & Dovichi, N.J. (2002), Fully automated two-dimensional capillary electrophoresis for high sensitivity protein analysis, Molecular & Cellular Proteomics, Vol.1, No.1, (January 2002), pp. 69-74, ISSN 1535-9476
40. Miller, J.M. (2005). Chromatography: concepts and contrasts, (2nd ed.) Wiley, ISBN 0471472077, Hoboken, USA
41. Mommers, J.; Knooren, J.; Mengerink, Y.; Wilbers, A.; Vreuls, R. & van der Wal, S. (2011). Retention time locking procedure for comprehensive two-dimensional gas chromatography. Journal of Chromatography A, Vol.1218, No.21 (May, 2011), pp. 3159-3165 ISSN 0021-9673

42. Nielsen, N-P.; Cartensen, J.M. & Smedsgaard, J. (1998). Aligning of single and multiple wavelength chromatographic profiles for chemometric data analysis using correlation optimised warping. Journal of Chromatography A, Vol.805, No.1-2 (May 1998), pp. 17-35 ISSN 0021-9673
43. Otto, M. (2007). Chemometrics, Wiley-VCH, ISBN 978-3-527-31418-8, Weinheim, Germany
44. Persson, P.O. & Strang, G. (2003). Smoothing by Savitzky-Golay and Legendre filters, In: Mathematical Systems Theory in Biology, Communications, Computation and Finance, Rosenthal J. Gilliam D.S., pp. 301-315, IMA Vol. Math. Appl., 134, Springer, ISBN 978-0387-40319-9, New York, USA
45. Pierce K.M.; Hope J.L.; Johnson K.J.; Wright B.W. & Synovec R.E. (2005). Classification of gasoline data obtained by gas chromatography using a piecewise alignment algorithm combined with feature selection and principal component analysis. Journal of Chromatography A, Vol.1096, No.1-2, (November 2005), pp. 101-110, ISSN 0021-9673
46. Poole, C.F. (2003). The Essence of Chromatography, (1st ed.), Elsevier, ISBN 0444501983, Amsterdam, The Netherlands
47. Rajalahti, T.; Arneberg, R.; Berven, F.S.; Myhr, K.M.; Ulvik, R.J. & Kvalheim, O.M. (2009a). Biomarker discovery in mass spectral profiles by means of selectivity ratio plot. Chemometrics and Intelligent Laboratory Systems, Vol. 95, No. 1, (January 2009), pp. 35-48, ISSN 0169-7439
48. Rajalahti, T.; Arneberg, R.; Kroksveen, A.C.; Berle, M.; Myhr, K.M. & Kvalheim, O.M. (2009b). Discriminating Variable Test and Selectivity Ratio Plot: Quantitative Tools for Interpretation and Variable (Biomarker) Selection in Complex Spectral or Chromatographic Profiles. Analytical Chemistry, Vol. 81, No. 7, (April 2009), pp. 2581-2590, ISSN 0169-7439
49. Reid, R.G.; Durham, D.G.; Boyle S.P.; Low, A.S. & Wangboonskul, J. (2007). Differentiation of opium and poppy straw using capillary electrophoresis and pattern recognition techniques. Analytica Chimica Acta, Vol.605, No. 1, (December 2007), pp. 20-27, ISSN 0003-2670
50. Ruiz-Samblas, C.; Cuadros-Rodriguez, L.; Gonzalez-Casado, A.; Rodriguez Garcia, F.D.P; de la Mata-Espinosa, P.; Bosque-Sendra, J.M. (2011). Multivariate analysis of HT/GC- (IT)MS chromatographic profiles of triacylglycerols for classification of olive oil varieties, Analytical and Bionalytical Chemistry, Vol.399, No.6 (February 2011), pp. 2093-2103, ISSN 1618-2642
51. Sarkar, S.; Dutta, P.K. & Roy, N.C. (1998). A blind-deconvolution approach for chromatographic and spectroscopic peak restoration, IEEE transactions on instrumentation and measurement, Vol.47, No.4 (August 1998), pp. 941-947, ISSN 0018-9456

52. Savorani, F.; Tomasi, G. & Engelsen, S.B. (2010). Icoshift: A versatile tool for the rapid alignment of 1D NMR spectra. Journal of Magnetic Resonance, Vol.202, No.2, (February 2010), pp. 190-202 ISSN 1090-7807
53. Sinkov, N.A. & Harynuk, J.J. (2011a). Cluster resolution: A metric for automated, objective and optimized feature selection in chemometric modeling. Talanta, Vol.83, No.4, (January 2011), pp. 1079-1087, ISSN 0039-9140
54. Sinkov, N.A.; Johnston, B.M.; Sandercock, P.M.L. & Harynuk, J.J. (2011b). Automated optimization and construction of chemometric models based on highly variable raw chromatographic data. Analytica Chimica Acta, Vol.697, No.1-2, (July 2011), pp. 8-15, ISSN 1873-4324
55. Skov, T.; Hoggard, J.C.; Bro, R. & Synovec, R.E. (2009). Handling within run retention time shifts in two-dimensional chromatography data using shift correction and modeling. Journal of Chromatography A, Vol.1216, No.18, (May 2009), pp. 4020-4029, ISSN 0021-9673
56. Stanimirova, I.; Boucon, C. & Walczak, B. (2011). Relating gas chromatographic profiles to sensory measurements describing the end products of the Maillard reaction. Talanta, Vol.83, No 4, (January 2011), pp. 1239-1246, ISSN 0039-9140
57. Tauler, R. & Barceló, D. (1993). Multivariate curve resolution applied to liquid chromatography-diode array detection. Trends in Analytical Chemistry, Vol.12, No.8, (1993), pp. 319-327, ISSN 0165-9936
58. Teofilo, R.F.; Martins, J.P.A. & Ferreira, M.M.C. (2009). Sorting variables by using informative vectors as a strategy for feature selection in multivariate regression. Journal of Chemometrics, Vol.23, No.1-2, (January-February 2009), pp. 32-48, ISSN 0886-9383
59. Tomasi, G.; Van den Berg, F. & Andersson, C. (2004). Correlation optimized warping and dynamic time warping as preprocessing methods for chromatographic data. Journal of Chemometrics, Vol.18, No.5, (May 2004), pp. 231-241, ISSN 0886-9383
60. Toppo, S.; Roveri, A.; Vitale, M.P.; Zaccarin, M.; Serain, E.; Apostolidis, E.; Gion, M., Mariorino, M. & Ursini, F. (2008). MPA: A multiple peak alignment algorithm to perform multiple comparisons of liquid-phase proteomic profiles. Proteomics, Vol.8, No.2, (January 2008), pp. 250-253 ISSN 1615-9861
61. Van den Berg, F.; Tomasi, G. & Viereck, N. (2005). Warping: investigation of NMR preprocessing and correction, In: Magnetic Resonance in Food Science: The Multivariate Challenge, Engelsen, S.B., Belton, P.S., Jakobsen, H.J., pp. 131-138, Royal Society of Chemistry, ISBN 0854046488, Cambridge, UK

REFERENCES

62. Van Nederkassel, A.M.; Dazykowski, M.; Eilers, P.H.C. & Vander Heyden, Y. (2006). A comparison of three algorithms for chromatograms alignment. Journal of Chromatography A, Vol.118, No.2 (June 2006), pp. 199-210 ISSN 0021-9673
63. Vivó-Truyols, G.; Torres-Lapasió, J.R.; Caballero R.D. & García-Alvarez-Coque, M.C. (2002). Peak deconvolution in one-dimensional chromatography using a two-way data approach. Journal of Chromatography A, Vol.958, No.1-2, (June, 2002), pp. 35-49, ISSN 0021-9673
64. Vosough, M.; Bayat, M. & Salemi, A. (2010). Matrix-free analysis of aflatoxins in pistachio nuts using parallel factor modeling of liquid chromatography diode-array detection data. Analytica Chimica Acta, Vol.663, No.1, (March 2010), pp. 11-18. ISSN 0003-2670
65. Watson, N.E.; VanWingerden, M.M.; Pierce, K.M.; Wright, B.W. & Synovec, R.E. (2006). Classification of high-speed gas chromatography-mass spectrometry data by principal component analysis coupled with piecewise alignment and feature selection. Journal of Chromatography A, Vol.1129, No.1, (September, 2006), pp. 111- 118, ISSN 0021-9673
66. Yao, W., Yin, X. & Hu Y. (2007). A new algorithm of piecewise automated beam search for peak alignment of chromatographic fingerprints. Journal of Chromatography A, Vol. 1160, No.1-2, (August 2007), pp. 254-262. ISSN 0021-9673
67. Yoshida H.; Leardi R.; Funatsu K. & Varmuza K. (2001) Feature selection by genetic algorithms for mass spectral classifiers. Analytica Chimica Acta, 446, 1-2, (November 2001), pp. 485-494, ISSN 0003-2670
68. Zhang D.; Huang, X.; Regnier, F.E. & Zhang, M. (2008). Two-dimensional correlation optimized warping algorithm for aligning GC×GC-MS data. Analytical Chemistry, Vol.80, No.8 (April 2008), pp. 2664-2671, ISSN 0003-2700
69. Zhang, Z.M.; Chen, S. & Liang, Y.Z. (2010). Baseline correction using adaptive iteratively reweighted penalized least squares. Analyst, Vol.5 (February 2010), pp. 1138-1146, ISSN 0003-2654
70. Zorzetti, B.M.; Shaver, J.M. & Harynuk, J.J. (2011). Estimation of the age of a weathered mixture of volatile organic compounds. Analytica Chimica Acta, Vol.694, No.1-2, (May 2011), pp. 31-37, ISSN 0003-2670

CHAPTER 10

Chemometrics of Cells and Tissues Using IR Spectroscopy – Relevance in Biomedical Research

*Ranjit Kumar Sahu and Shaul Mordechai**

Department of Physics, Ben-Gurion University of the Negev, Beer-Sheva
Cancer Research Center, Ben-Gurion University of the Negev, Beer-Sheva
Israel

1. INTRODUCTION

Biochemical analyses of substances rely upon the ability of techniques to identify qualitatively and quantitatively the components present and are based on physicochemical characteristics as well as chemical nature of substances being detected. While chemical analyses usually depend on reactions of a given substance and can be destructive, spectral studies are usually non-destructive and deal with describing a substance based on properties like absorption or transmission of light (e.g. UV, Visible, Infrared (IR)), light scattering ability, fluorescence /phosphorescence using various optical techniques. Thus, the technique (Fourier transform infrared) spectroscopy has gained prominence in both research and applications in different fields of science. Among the various techniques, IR spectroscopy owing to its lower potency of causing damage compared to X-rays, gamma rays and UV rays (as it is based on weak vibrational energies) has become the technique of choice during chemical analysis of substances. IR spectroscopy can not only provide information about the various components in a complex material but is also unique in its ability to be modified into different kinds of instrumentations based on requirement. The various IR spectroscopy based instruments from a simple IR based spectroscope that helps to obtain the absorbance spectra of a chemical compound to the complex imaging systems that employ computational methods in addition to the technical sophistication are based on a simple principle that every compound or a particular combination of compounds can be described by means of a FTIR (Fourier transform infra red) spectra qualitatively and quantitatively.

The guiding principle of all such analyses lies in the fact that when IR radiation of different wave numbers are simultaneously passed through a sample, specific wave numbers are absorbed based on the vibrations of molecules, creating a unique fingerprint of each sample, from a simple molecule like a protein molecule to a more complex structure like eukaryotic tissues. In spite of the fact that the cells and tissues can be discriminated based on their spectral fingerprints in the mid IR/NIR region, their signatures are the result of contribution from several biological components that at times absorb at similar or overlapping wave numbers. In order to explain the contribution of different metabolites like carbohydrates, nucleic acids, proteins and lipids in a sample, spectra of pure components are collected and analyzed for specific patterns. Comparison of compound that vary in one or more functional groups aids in the determination and assignment of particular groups and defines the specificity of signature for each along the entire spectral region. Assignment of the exact contribution of each component to the entire spectra therefore makes the method quantitative as well as qualitative.

Figure 1A shows representative spectra of blood fractions obtained by FTIR-MSP (Fourier Transform Infrared Microscopy), mounted on ZnSe slides in transmission mode. It is observed that in spite of the diversity of the source and type of the samples, absorbance of IR occurs at similar wave numbers, implicating that the samples are composed of similar basic substances as mentioned above. Simultaneously it also shows how the variation occurs among these substances and what are the likely principal components of each, contributing to its unique spectra. For example, plasma has less sugar compared to the bacterial cell which has a capsule and thus has more prominent peaks in the region 900-1185. Similarly, the RBCs that lack nucleic acids show diminished peaks while WBCs show clear absorbance peaks between 1185 cm^{-1} and 1300 cm^{-1}.

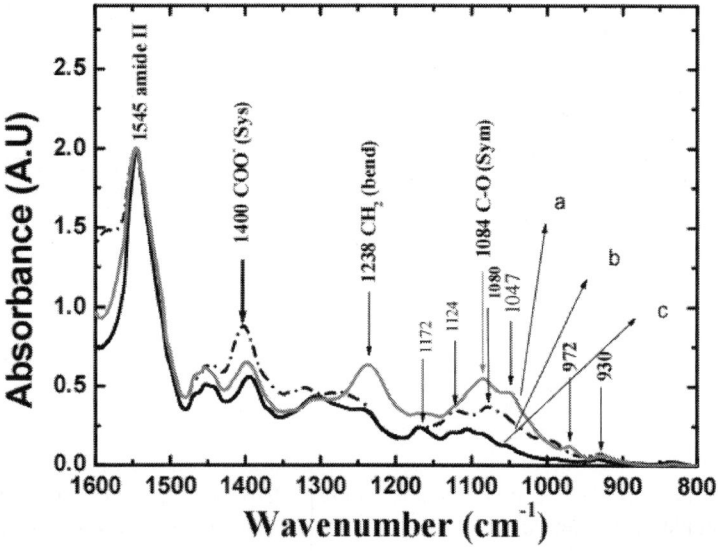

Figure 1A. Amide II normalized average spectra in the region 800-1800 cm-1 of different ZnSe mounted blood components (a-WBC, b-Plasma and c-RBC) analyzed by FTIR-MSP in transmission mode showing few peaks absorbance and the functional groups associated with it.

1. INTRODUCTION

Thus a qualitative assessment can be made regarding the composition of a sample by looking at its spectra and also defining what type of functional groups could be contributing to absorbance at specific wave numbers.

Several tissues and organisms can show common peaks due to similarity in composition. As shown in Figure 1b, there is absorbance at wave numbers corresponding to components (like carbohydrates, proteins) that are present across a wide range of samples. Owing to large number of data due to the availability of several characteristic wave numbers for individual compounds, mathematical and computational methods are developed that can analyze spectra as per the requirement, providing users the convenience of obtaining the data in usable and interpretable forms. Advancements in computational techniques have added to the utility of the FTIR based instruments by making spectral calculations rapid and automatic, leading to their application in diverse fields of chemistry and biology for both applied and basic research. The potential of FTIR spectroscopy thus serves not only for routine applications but also as a diagnostic tool where other optical methods become difficult to apply. In view of these developments an ever expanding field of biomedical research based on FTIR based technologies has arisen over the last few decades. With its unique abilities, the technique has been applied mostly in cancer diagnosis and monitoring, microbial identification and drug efficacy evaluations to name a few. The present chapter describes in brief the different aspects of applications of FTIR in biomedicine and their suitability and relevance to biomedical research.

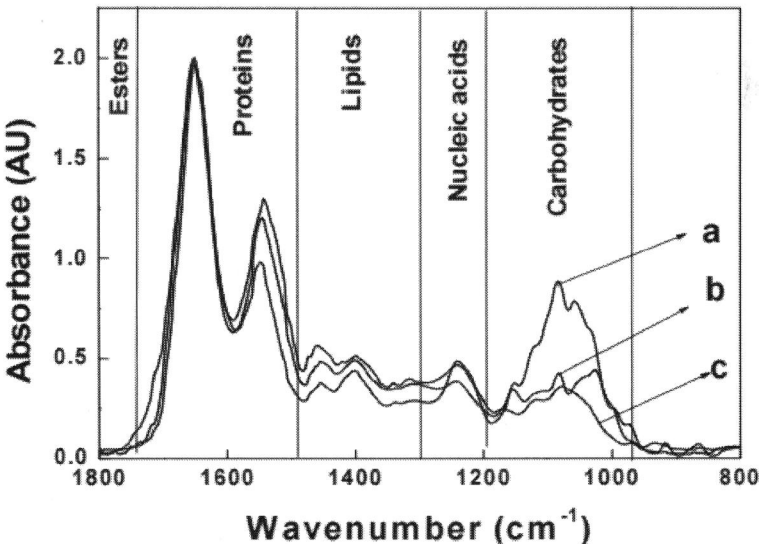

Figure 1B. Amide I normalized, averaged spectra of (a) bacteria (Streptococcus pneumoniae), (b) Cervical epithelium intermediate layer and (c) colonic crypt in transverse section to show absorbance of different biological components at different regions of mid IR. Note that though the carbohydrates absorbance occurs principally in the region between 1200-900 cm-1, the pattern differs across samples, giving a preliminary information of their different origins.

2. BRIEF HISTORICAL PERSPECTIVE OF FTIR SPECTROSCOPY IN BIOMEDICAL RESEARCH

IR spectroscopy has been in use for chemical analysis for several decades. The recognition of its ability to distinguish between normal and abnormal based on fingerprints of the cells and tissues has been utilized to describe changes at molecular and cellular levels (Yang et al 1995, Malik et al 1996, Schultz et al 1996, Malins et al 1997) opened up a new area in the field of biodiagnostics. Other studies paved way for the examination of the technique as a diagnostic tool for identification of disease especially malignancy (Cohenford et al 1997, Rigas et al 1990, Wang et al 1997). This was followed by expanding the methods to study tissues from several different organs such as skin, cervix, liver among others with an emphasis on cancer identification which would be at par with conventional diagnostic techniques without the negative factors like toxicity to live tissues. Early works focused on defining the tissue variability in terms of IR spectra (Chiriboga et al 1998a,b,c, Wood et al 2004). This led to a large number of studies with different tissues and cell lines with an objective to make the technique relevant to oncology (Fukuyama et al 1999, Diem et al 2000, Gao et al 1999, Yano et al 2000, Malins et al 2003). The potential application in other biomedical fields was expanded by studying the classification of microorganisms using the technique and using it for identifying pathogens (Kirschner et al 2001, Choosmith et al 2001, Sandt et al 2003, Essendoubi et al 2005). These studies led to the development of the concept of "biomarkers" which are parameters or statistics derived from the spectral data that help to identify or differentiate among samples. The most promising biomarkers have been repeatedly used though developments have led to utilization of more complex spectral information rather than simple ratios. For example the region between 900-1300 cm^{-1}. There were also studies undertaken to identify universally suitable biomarkers which could be used in different cells or tissue types to identify malignancies (Sahu et al 2004b, 2005, Mordechai et al 2004). Thus, there was a parallel development of mathematical and computational methodologies to utilize the spectral data and improve the sensitivity and specificity of the diagnosis and helping clearly defining ambiguous samples or accounting for outliers. These methods include linear discriminant analysis (LDA) (Krafft et al 2006), probabilistic neural networks (PNN) (Mark et al 2004), Principal component analysis(PCA) (Muralikrishna et al 2005) artificial neural networks (ANN) (Zhang et al 2003, Lasch et al 2007) and Discriminant classification functions (DCF) (Sahu et al 2010, Bogomolny et al 2007). The methods like clustering of spectra based on either Wards algorithm using Euclidean distances or Mahalanobis distances have been used to classify tissues and used as a means of developing pseudocolor images during FPA mapping of tissues (Lasch and Naumann, 1998). Similarly while the diagnosis of tissues using FTIR was continuing, the detection of changes in single cells using more sophisticated techniques like synchrotron or SERS were undertaken (Chekun et al 2002). Studies were undertaken to elucidate the contribution of individual components like the nucleic acids which held a great promise as biomarkers (Malins et al 2005, Sahu et al 2008). The works were also supported by inducing transformation in cells using various biological and chemical reagents and studying the spectral manifestations (Ramesh et al 2001, Salman et al 2003, Bogomolny et al 2008). The other significant technical development undertaken was the fiber optic systems that could be used for surface scanning or as probes and these are based on the ATR systems. Parallel research was also carried out to help identify and exclude contaminants in samples (Wood et al 1998, Romeo et al 2003, Sahu et al 2005) or use samples from paraffin embedded systems (Ly et al 2008). The effect of physical phenomenon like Mie scattering and its contribution to the FTIR data was also studied (Kohler et al 2008, Lee et al 2007) with an aim to understand how it can interfere with and influence the basic spectral data. Thus in the last two decades progress has been made in several directions to help realize the full potential of the technique in various biomedical fields making the data both qualitatively and quantitatively relevant. Table I lists a few studies where detection of cancers was carried out using FTIR based instrumentation.

3. DIFFERENT TYPES OF FTIR SPECTROSCOPY AND THEIR SUITABILITY TO DIFFERENT FIELDS OF BIOMEDICAL RESEARCH

Currently the simplest and most familiar IR spectrometers existing consist of a source of IR beam, a sample holder and a detection system to monitor the absorbance. Utilizing several different matrices that are IR transparent (KBr, BaF2, CaF2, ZnSe), it is possible to obtain the spectra from samples as diverse as pure compounds such as proteins, lipids, drugs and metabolites to homogeneous preparations such as body fluids, microorganisms, cell lines that can be mounted without destroying their chemical nature. While it is imperative that each sample would require a different approach before being mounted for analysis, the common requirement is removal of water (that interferes with the spectra) through processes like lyophilization or even heating thermo stable samples in an oven. The samples are at times dissolved in D_2O to over come the effect of water especially where they are aqueous soluble. Such systems are used to study interactions between biological molecules such as proteins and nucleic acids with ligands like toxins and antibiotics (Pratibha and Malathi, 2008, Bourassa et al 2011).

A slightly different approach is required for cells and microbes where a homogeneous material is available (e.g cell lines, bacteria). Such samples require removal of extraneous materials that can confound the spectra and are usually subjected to washes by normal saline before being dried for spectral acquisition. For example, bacterial cells from cultures in log phase are harvested after treatment with chloroform and the media removed by centrifugation. Subsequently they are washed several times in normal saline (without phosphates) and mounted on discs or any IR transparent support for spectra collection (Sahu et al 2006b).

Table 1. Malignancy in different types of tissue studied by FTIR spectroscopy.

Authors	Tissue/Organs	Analysis	Region/Wave number
Yang et al 1995	Fibrosarcoma	Intensity	1085
Fujioka et al 2004	Gastric cancer	Discriminant Analysis	925-1660
Podshyvavlov et al 2005	Cervical cancer	PNN	900-1800
Krishna et al 2007	Ovarian tissue	Cluster analysis	1540-60 & 1720-1780
Ali et al 2008	Brain tumor	Cluster analysis	Higher region, 1735.
Maziak et al 2007	Oesophagus	Ratios	Several
Harvey et al 2007	Prostate	FTIR-acoustics	
Argov et al 2002	Colon	ANN	900-1800
Yano et al 2000	Lung	Ratios	1045,1467
Wu et al 2001	Oral tissues	Intensity	1745
Lasch and Naumann 1998	Melanoma	ANN/cluster analysis	
Andrus and Strickland 1998	Lymphoma	Ratio	1020,1121

When dealing with eukaryotic cells like cell lines, the cells are harvested and washed in a buffered solution several times to get rid of the media components and finally washed in normal saline before removing the excess water by a process of air drying. Similarly, different components of blood can be separated and fractions of plasma directly dried on the support. For the cellular components of blood or fluid biopsies like fine needle extracts by different biophysical techniques can be used for obtaining a given population of cells such as RBC, WBCs, monocytes or T cells and B cells which after purification are treated similar to cell lines for spectroscopic measurements. Most of these samples can be measured with routine spectrometers.

The approach becomes more complex when dealing with spectroscopy of tissues. In case of a relatively homogeneous tissue like liver, spectra can more or less be obtained from any region of the biopsy while tissues like cervical tissues that have clearly differentiated zones would need measurements at precise locations to alleviate spectral variation due to location effects (Chiriboga et al 1998a). However these samples necessitate a microscopic evaluation before measurement. Colonic biopsies that display a more or less uniform pattern in the cross section are measured on the circles of crypts for diagnosis of malignancies. However the longitudinal sections of the crypt require a more defined location for measurement and invariably depend on the utilization of microscope. More complicated measurements of tissues where several different kind of cells are required to be measured depend on focal plane array detectors where each pixel of the measured area can be represented by a spectra. The requirement of methodologies for reconstruction of a pseudo image based on spectral characteristics become the norm and essential feature of such measurements.

4. FTIR INSTRUMENTATION AND THEIR APPLICATIONS IN BIOMEDICINE

The simplest FTIR spectrometers have been most widely used to study cells and plasma samples. When a sample's spectra are obtained, they are normalized, averaged and baseline corrected. The usual practice is to undertake a baseline correction in the spectral region of interest using the rubber band form and locating the two extreme points of the region of interest. The cut spectra are then normalized again to the highest peak or the area. The different methods of baseline correction in specified regions would greatly alter the results. Hence application of similar data processing on all spectra being considered in a study becomes essential. The intensities at selected wave numbers are then used to define ratios or biomarkers that can define the criterion being sought for the diagnosis. Extracting intensity ratios is preferred than band intensity, since it yields a dimensionless quantity which is mostly independent on the exact normalization procedure used (such as Min-Max or vector normalization). Other than the intensities, many studies report shifts in peak wave numbers which can provide additional valuable information. However instruments that can measure only a few wave numbers can also be used when more sophisticated instrumentations are not available. For example measuring the band intensities of the CH_2 and CH_3 vibrations between 2800-3200 cm^{-1} can provide information about the status of the tissue (Sahu et al 2006a,2005). Similarly various other parameters have been routinely used in diagnosis of diseases and cancer. The ratio of Amide I /Amide II quantified using the integrated absorption of Amide I and Amide II (1750-1590 cm-1, 1590-1480 cm-1) was one of the first ratios found suitable for diagnosis and is an indicator for the DNA absorbance variation (Liquier and Talliander,1996, Benedetti et al., 1997; Gasparri & Muzio, 2003). RBCs (red blood cells) from humans that lack nuclei posses a ratio of unity for this parameter and maybe used as a reference when understanding nuclear changes using these wave number regions.

Similarly quantification of phosphate metabolites calculated by measuring the integrated area of phosphate symmetric (990-1145 cm^{-1}) and antisymmetric (1190-1275 cm-1) bands have been used to understand the stages of the cell (Yang at al. 1995). The most promising phosphate absorbance that has been used in many studies is the RNA/DNA ratio dependent on the phosphate absorbance arising from the symmetric vibrations of DNA and RNA namely at 1020 and 1121 cm^{-1} respectively. Several other band intensities have been used individually or in combination as listed in table 1 for evaluating cells, tissues and biopsies. The band at 1045 cm^{-1} is attributed to the vibrational frequency of -CH_2OH groups and the C-O stretching frequencies coupled with C-O bending frequencies of the C-OH groups of carbohydrates (including glucose, glycogen, etc.) as well as the capsular carbohydrates of bacteria. The ratio of the integrated absorbance at 1045 cm^{-1} to that at 1545 cm^{-1} provides an estimate of the carbohydrate absorption (Parker, 1971). When a normalization is made to the amide II band the intensities at this wavenumber can be assumed to directly correlate with the amount of carbohydrates and is similar to normalizing the amount to protein in cells and tissues using chemical analyses.

5. PROTOCOLS FOR SAMPLE PREPARATION

As the sources of samples for FTIR spectroscopy can be different, they need special preparations adopting different kinds of procedures depending on the samples.

5.1 FTIR Spectroscopy of Homogeneous Materials

5.1.1 Drug macromolecules interactions

Studies of drug interactions by FTIR spectroscopy are easily carried out in systems using D_2O. Various approaches are possible. The drug and protein/nucleic acid interaction is carried out under solution conditions at predetermined stoichiometry and the mixture is then added to a film from which spectral data are obtained. The individual components are also used as references or controls. The difference spectra between the control and with the ligands is sued to quantify the amount of bound ligands or study the changes in the protein secondary structure by monitoring the shifts in peak intensities or variation in peak intensities of different functional groups (Bourasssa et al 2011). FTIR-ATR system can also be used for studying the interactions using a dialysis system. (Krasteva et al 2006, Kumar & Barth (2011).

5.1.2 Microbial cultures

Bacterial cultures growing in the log phase are fixed with formaldehyde (final concentration 0.25%) and washed three times with saline containing 0.25% formaldehyde and once with saline only. The pellet is resuspended with saline to an $OD_{450} = 0.3$ (about 20 ml). One microliter of the suspension is spotted on Zinc-Selenium slides and air dried for 6 hours to remove any water in the sample under a laminar air flow chamber. Microscopic FTIR measurements are made in transmission mode using the FTIR microscope IRscope II with a mercury-cadmium-telluride (MCT) detector, coupled to the FTIR spectrometer (Bruker Equinox model 55/S, OPUS software). Absorbance is measured from an area of 100 nm diameter by setting the slit to 100 microns. Regions of thickness of about 10 microns (as seen from the ADC values) are selected and 128 scans co added for each spectrum. For each sample at least five spots are measured and the average spectra calculated. Similar procedure can be adopted for yeast cells. In case of fungi which have a tendency to grow as filaments (hyphae), the filaments are harvested and briefly tweezed or torn to get uniform untangled mass as observed under the microscope. These are then washed in water or saline before mounting on the slides. These can also be measured using ATR systems as they are highly adherent and a spread sample is likely to have non uniformity in thickness owing to the clumping tendency of the filaments (Salman et al 2011).

5.1.3 Cell lines and transformed cultures

Cell lines are cultured under suitable conditions (e.g. 37°C in RPMI medium supplemented with 10% of newborn calf serum (NBCS) and the antibiotics penicillin, streptomycin) and passed to obtain confluent cells. Adherent cells are harvested using a cell lifter while cells in suspension are taken directly. The cell suspensions is centrifuged for 5 minutes at 3000 rpm and the pellet washed in phosphate buffered saline several times (3-5 times). After the washing, the cells are washed at least two times in normal saline and resuspended in saline such that the cells density is about 1×10^6/ml. One microlitre of this suspension is spotted on a ZnSe window and allowed to air dry in a laminar air flow chamber for several hours. A microlitre of the last supernatant before suspension may also be spotted to be used as a reference for any unwanted materials. Spectra from this sample should ideally be similar to the background spectra. Cells and microbes like fungus may also be grown on IR transparent matrices when they are analyzed by ATR or in reflectance mode. Precautions are however required to avoid artifacts due to very thin layers of drying of cells (Mourant et al 2003a).

5.1.4 Samples from body fluids such as blood or urine

Body fluids usually contain cells and non cellular components, tissue debris and a fluid or matrix. When the cells are the material of interest, they are isolated into pure or homogeneous forms using various separation techniques that utilize the different sedimentation coefficient of the cells and including Fluorescence activated cell sorter (FACS) when required to isolate a pure population based on their cell surface markers. In case of blood samples, the collection is made in heparin or EDTA to prevent the clotting. Then they are separated by a histopaque or Ficoll gradient. The cells are then washed in PBS followed by normal saline as for cultured cells. For spectra of other components such as plasma from blood, the plasma or sera obtained from blood is directly spotted on the ZnSe windows without further processing. However, if the liquid is very viscous, it can lead to super saturated spectra. To avoid this, the fluid is spread by dragging the drop in several directions with the tip and spreading it. Study of other body fluids like urine may require filtration of the liquid to remove debris and its concentration to reduce the amount of water before it is loaded on matrices. In case of study of the cellular components, a prior separation of different cell components maybe required.

5.2 FTIR spectroscopy of tissue samples

Most of the diagnostic potential of FTIR spectroscopy has been evaluated in tissues that relate to carcinogenesis with an objective to diagnose malignancy or premalignancy. Thus, conventionally the material available was formalin fixed, paraffin embedded tissues. In most of these cases, the tissue is sliced to a suitable thickness(usually 10 microns), processed to remove the embedding material and mounted on a slide (usually a ZnSe slide) before it is used for measurements. Such slides are stored in histology frames dipped in 70% ethanol to prevent contamination. A consecutive section that is congruent to the section being measured is usually used for H&E staining. Thus, both the optical as well as the FTIR measuring capabilties of the microscope are used to pin point the exact location of measurement. For example, in tissues like cervical epithelium, where there is a decrease in glycogen level from the superficial layer to the basal layer (Chiriboga et al 1998a), and also the reduction of glycogen is an important marker for CIN or cervical cancer, it becomes important to measure at defined regions, namely the intermediate layers of the tissue to accurately be able to classify the stage of CIN (Figure 3a) (Mark et al 2004). In case of colonic epithelium, the tissue can contain regions where the crypts are seen in a transverse or longitudinal section. The measurements can be made on either type of section. However, the transverse sections are usually measured for conventional diagnosis and grading for cancer or premalignancy (Argov et al 2002, Argov et al 2004). Measurements along the crypt height have been employed to understand metabolic activity in terms of spectral changes (Salman et al 2004) or to define abnormal proliferation in the epithelium (Sahu et al 2004a, Sahu et al 2010). In these cases, a region of the slide free from contaminants is used for the background. The histology of the tissue is evaluated on the complementary slide which is stained with H&E. Regions are selected that indicate the required stage of disease and then measured by locating the identical spots on the ZnSe slide. At least five measurements on each sample are carried out. The morphology of the colonic epithelium is well defined and it is easy to locate the crypts even while viewing under the optical microscope. Thus, it is easier to locate the crypts and measure them. Usually in these tissues, measurements are made by adjusting the slit to 120 microns that results in an area of 100 micron diameter. As most formalin fixed tissues are paraffin embedded and processed, the spectroscopy in the higher region can also be used to monitor contaminants like blood and paraffin or any residual cleansing reagent like xylol (Sahu et al 2005). Both the transverse sections of the crypt, where the entire crypt would fit into the area of measurement or several locations along the crypt in the longitudinal section can be measured based on the requirement of the studies (Argov et al 2002, Argov et al 2004, Sahu et al 2004a, Salman et al 2004). In case of measurement along the length of the crypt the replication is carried out by measuring several crypts at similar distance form the base or apex unlike the cross section where several crypt circles are measured. For more complex tissues such as skin, where different kinds of abnormalities occur (both benign and malignant) the exact location of the sample measurement area requires the involvement of an expert dermatologist/pathologist to pin point and mark the areas on the complementary slides (Hammody et al 2007). Advancement in technologies have made it possible to automatically measure entire regions of the biopsy using the FPA detector systems and the spectra corresponding to every pixel of the measured area can be obtained. This type of measurement is slow though automated and helps to map entire regions. However the study on a complementary slide to demarcate the exact location of histological entities is still essential to precisely correlate the spectra. The processing is similar to microscopy but owing to a large amount of acquired data, automated computational methods become necessary.

5.3 Cellular Transformations

Cellular transformations are used to understand spectral changes happening due to biological changes in cells and tissues, often inducing controlled changes in cell lines using genes or viruses. Cells grown in tissue culture plates are treated with polybrene (a cationic polymer required for neutralizing the negative charge of the cell membrane) for 24 h before infection with the virus. Free polybrene is removed and incubated with the high titer infecting virus stock at 37°C for 2 h. The unabsorbed virus particles are removed by washing the cells in fresh warm medium and fresh medium is added. Several wells are used in a study including a control group passed through similar conditions without the infecting virus. After various stages of progression the cells are examined for the appearance of malignant transformation and cells from sets of wells harvested at different time intervals used for spectra acquisition. Cells can also be grown on soft agar or stained to confirm their malignant transformation.

6. SPECTRA ACQUISITION AND DATA PROCESSING

Spectra can be obtained either in transmission or reflective mode or both from a FTIR microscope (Argov et al 2002, Chang et al 2003). Samples loaded on a mounting material such as a ZnSe slide are observed under the microscope for their uniformity and thickness and representative regions are selected. Prior to measuring the samples a background measurement is made in a region free of any samples or reference material. This spectrum is saved as a background (Figure 2a). The data processor automatically subtracts these background spectra when further measurements are made. The reference samples if any are then measured. The thickness of the sample is usually reflected in the ADC rates. A ADC value of less than 3000 usually denotes a thickness of less than 10 microns and often leads to noisy spectra. On the other hand regions with a very high ADC value of (>6000) can give rise to supersaturated spectra. Once a set of reference spectra are obtained and they display required characteristics, the samples are measured. Ideally 3-5 spots are measured on each spotted sample.

The background corrected spectra is then baseline corrected for the entire region using a rubber band baseline function (Figure 2A). Selected spectral regions of interest are then separately cut from the entire spectrum and a second baseline correction is made (e.g region between 800-1800 cm^{-1}). This is followed by a normalization of the spectra (Fig.2B,C). The different spectra are then averaged and the average spectra used to represent a sample. Intensities at different wavenumbers obtained form this processed spectra are the inputs for various mathematical analyses. For example, the ratio of intensities at 1545 and 1045 cm-1 (Figure 2C) or peak areas after deconvolution (Figure 2D).

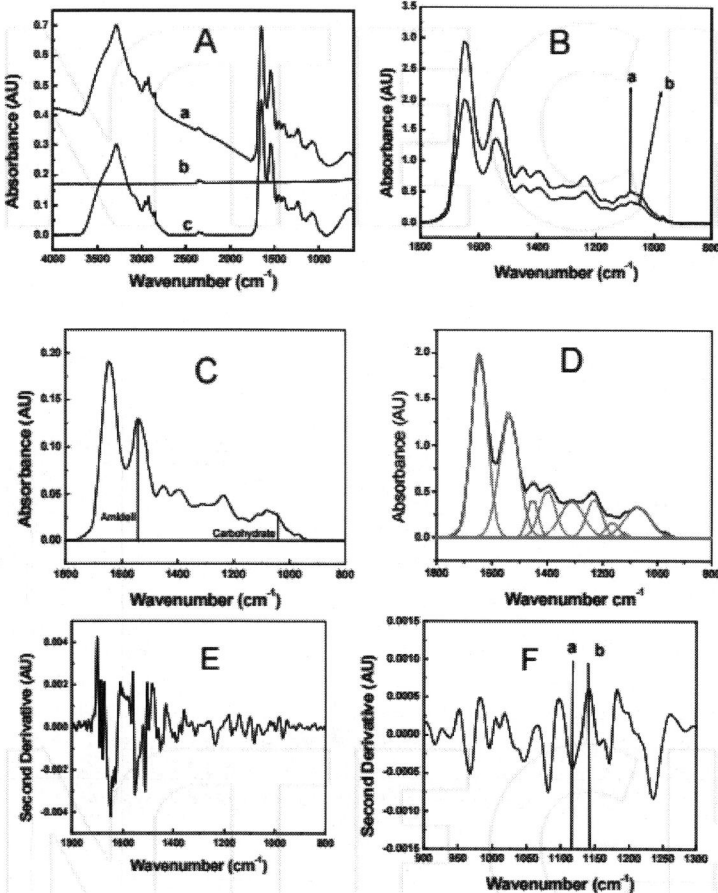

A. Initial absorbance spectrum collected in transmission mode (a), background (b) and background subtracted and rubber band baseline corrected spectrum (c). B. Spectra in the region 800-1800 cm-1 after cutting from the entire spectrum, baseline correction and normalization to amide I peak (a) or amide II peak (b). C. Area normalization of the spectrum after cutting the region between 800-1800 cm-1 preceded by baseline correction and followed by offset correction. The vertical lines denote intensities at 1545 and 1045 cm-1 which can be used for quantification of carbohydrates. D. Deconvolution of the spectra to obtain intensity at various peaks using Gaussian fit of the spectra. (E.) Second derivative of the spectra. (F) Selected region in the second derivative spectra where the intensity at a selected wavenumber is calculated by taking the intesities at the minimum (a) and the adjacent maximum (b).

Figure 2. Spectral acquisition and processing.

Usually a region with normal morphology is taken as a reference or control while identifying the malignancy in a tissue to avoid any heterogeneity due to individual variations while comparing normal and abnormal tissues (Argov et al 2002). With recognition of the fact that normal tissues may still display abnormal FTIR spectra it becomes important to precisely define a normal tissue or biological entity which in itself could lead to better understand dynamics of spectral variation during disease progression (Sahu et al 2010).

Second derivatives and higher order derivatives of spectra are used to avoid the errors creeping in due to bias originating from baseline selection methods (Figure 2E). In this case when the intensities are measured, the minima correspond to the maxima (peak intensity) in the original spectra. Sum of the value at the minima with the value at the nearest maxima is taken to indicate the intensity at the particular wavenumber (Fig 2F).

6.1 Spectral Acquisition Using ATR/FEWS Systems

ATR measurements can be done using Bruker FT-IR Tensor 27 Spectrometer equipped with a liquid nitrogen cooled mercury-cadmium-telluride (MCT) detector and coupled with Horizontal Attenuated Total Reflectance Accessory (HATR (Horizontal Attenuated Total Reflectance), PIKE technologies Inc,) systems (Bogomolny et al 2009). The accessory is connected with a nitrogen reservoir, which enables to preserve dryness of the sample and maintaining an inert atmosphere. The design employs a pair of transfer optics to direct the infrared beam to one end of IR transmitting crystal. A similar pair of flat mirrors directs the beam from the other end of the ATR crystal to the spectrometer detector. The ATR crystal made up of ZnSe is of a trapezoid shape with its thickness suitable to produce optimum performance. To reduce variance each sample is measured several times (at least five) and the ADC rates are empirically chosen between 4000-5000 counts/sec.

7. ADVANCES IN DATA ANALYSIS

Unique spectral fingerprints of biological entities in the mid IR region are the manifestation of several components absorbing at different wavenumbers, with overlap. Thus, both the quantity and type of individual components can alter the fingerprint. As shown in Figure 3, during carcinogenesis, in both colon and cervical tissues, there is a depletion of carbohydrates likely due to increased metabolism. However the disappearance of carbohydrates evident from vanishing of the triads between 900-1185 cm^{-1} in cervical tissues is slightly different than in colonic tissues. This is likely because glycogen is known to accumulate in cervical tissues, increasing in concentration from the basal to superficial layer, the absorbance associated with colonic tissues is more likely to be from glycoproteins than pure glycogen. Thus, though similar functional groups may contribute to the absorbance, they can manifest as spectral variations. This also necessitates that each tissue or cell type is investigated independently though common biomarkers are used. Such type of differences and the gradual variation at specific wave numbers due to transition of tissues or cells from one type to another requires that contribution of individual metabolites like carbohydrates, nucleic acids, proteins is clearly evaluated.

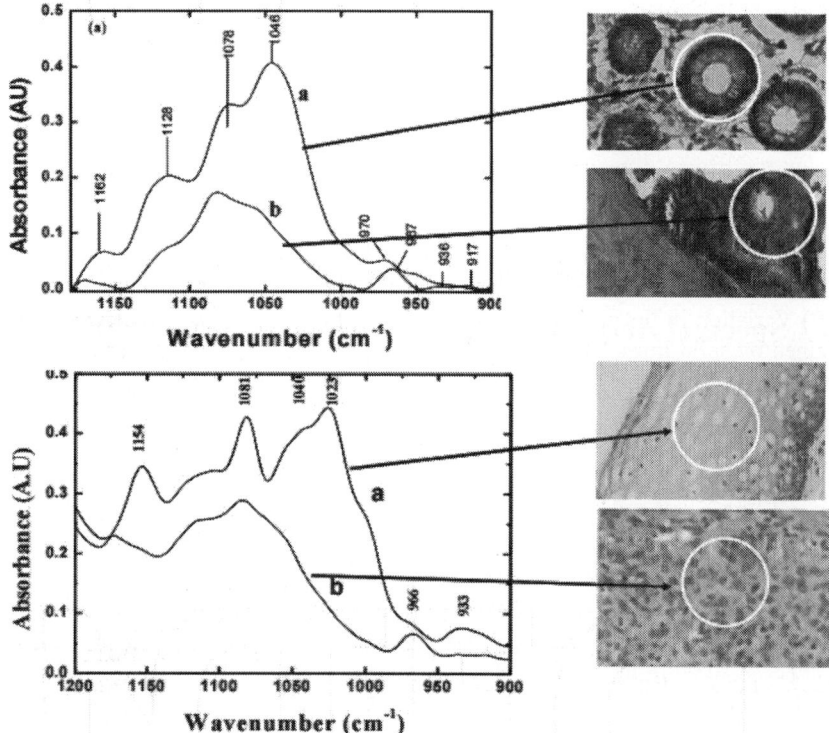

Figure 3. Baseline corrected spectra from (top) normal and cancerous colonic tissues and (bottom) normal and cancerous cervical tissues, indicating changes in the region 900-1200 cm-1. The circles in the histological sections depict the measurement sites. Note that spectra in the upper panel are further baseline corrected by using a rubber band baseline in the region 900-1200 cm-1 while in the lower panel the data presented are after normalization to the amide II peak (not shown). Note also the similarity in the trends in peak intensity on carcinogenesis irrespective of tissue origin.

Figure 4a shows the variation in the integrated intensity in colonic biopsies with different diagnostic features. It is noted that while both normal and hyperplastic samples show similar quantitative traits, the other biopsies with varying degrees of malignancy or premalignancy display a decrease in the levels. It is also noted that the samples with worst prognosis show similar values (For example C2 grade cancer and Severe level of hyperplasia have similar value). This indicates that systems with similar spectral features pertain to a class of conditions with similar outcomes. The feature can also be used to monitor the progression of the disease over time, indicating its possible potential as a biomarker (Figure 4b).

7. ADVANCES IN DATA ANALYSIS

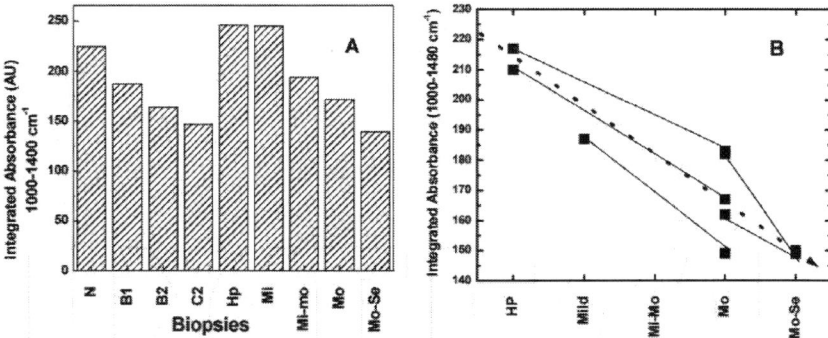

Figure 4. (A) Variation in levels of integrated absorbance between 1000-1480 cm-1 of colonic biopsies with different levels of malignancy or premalignancy. (B) Variation in the parameter over time. The biopsies of the same patients are joined by solid lines. The dotted arrow points to the progression of the disease. N=normal, C1, B2, C2=Cancer grades, Hp=hyperplastic, Mi=mild, Mo=moderate, Se=severe.

However the two groups of tissues need separate independent analysis i,e the samples of cancerous biopsies and hyperplastic biopsies when dealing from a clinical perspective. To further be able to differentiate between these two subgroups additional biomarkers become necessary. The availability of several biomarkers, suitable for disease diagnosis can complicate the selection though they can increase the sensitivity of the technique. Therefore other than using simple ratios, cluster analysis using several biomarkers or spectral data in entire regions can be undertaken. These help to minimize the false negatives or positives. Inclusion of mathematical and computational methods to analyze suitable wavenumbers for diagnosis focused on the differences between normal and abnormal tissues can utilize artificial neural networks and set up data bases that are used as a reference. Usually a part of the study sample is used to train the system before the blind samples are analyzed. Setting up of reference data base with spectra from clearly identified histopathological systems is a primary step in setting up of a good diagnostic software.

The potential of FTIR increased several fold by the combination of computational methods can thus also be used to overcome spatial and temporal variation in samples. The setting up of such database becomes more crucial when microorganisms (that have a tendency to mutate and change rapidly) are being studied as the older database can be used to monitor such evolution by studying spectral variations. Cluster analysis where different groups are separated by the distance proportional to their heterogeneity is another way to study the evolutionary relation between species and subspecies of microorganism. However clustering may not be sensitive enough to discriminate among closely related samples. Thus, more methods of analyses are resorted to like ANNs (Goodacre et al 1996). Artificial neural networks make it possible to examine samples over time by setting up reference data bases. These systems often work on the principal of classifying the tissue in a binary progression mode as depicted in Figure 5a. The final results are displayed as a confusion matrix where the probability of classifying a particular biopsy into one of the diagnostic group is expressed as a percentage.

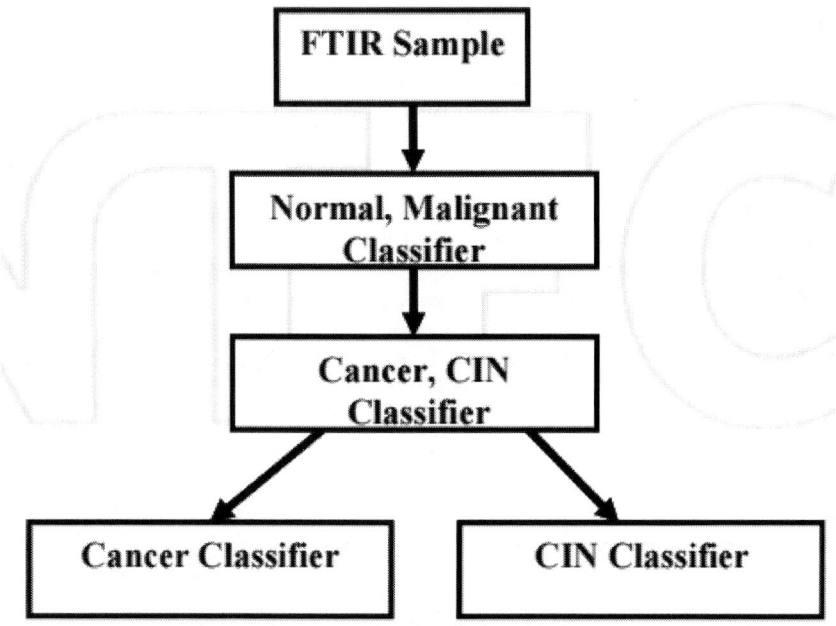

Figure 5. A model displaying a possible methodology to classify cervical biopsies using an ANN.

Data analysis involving FTIR spectra focuses on utilization of intensities or integrated absorbance from wave number/ wave number regions or their various combinations that result in separation of the different classes of samples under study. Often diagnosis between normal and abnormal tissues is carried out by monitoring absorbance at selected wave numbers after routine mathematical manipulation of the spectra. (Sahu et al 2004, Mordechai et al 2004). Later entire regions of spectra or their derivatives were used to classify tissues or samples (supervised or unsupervised) into clusters to determine their hierarchy or their relation with one another. These types of data conventionally presented as clusters were later used in advanced methods like FPA, to organize areas with similar into pseudocolor maps to establish patterns of tissues from FTIR spectra.

Cluster analysis based on the Ward's algorithm separates samples by the distance proportional to their heterogeneity and difference spectra of colonic crypts has successfully been used their classification (Sahu et al 2004). Similarly the cluster analysis has been used for classification of microorganisms (Sahu et al 2006). Figure 5b displays a schematic diagram of how different biopsies are classified and the closely related conditions tend to group together. Cluster analysis of FEWS spectra of human skin samples using a chemical factor

REFERENCES

7. Bombalska, A.; Mularczyk-Oliwa, M.; Kwaśny, M.; Włodarski, M.; Kaliszewski, M.; Kopczyński, K.; Szpakowska, M. & Trafny, E .A. (2010). Classification of the biological material with use of FTIR spectroscopy and statistical analysis. Spectrochim Acta A Mol Biomol Spectrosc. 2011 Apr;78(4):1221-6.
8. Bourassa, P.; Dubeau, S.; Maharvi, G.M.; Fauq, A.H.; Thomas, T.J. & Tajmir-Riahi, H.A. (2011). Locating the binding sites of anticancer tamoxifen and its metabolites 4- hydroxytamoxifen and endoxifen on bovine serum albumin. Eur J Med Chem. 2011 Sep;46(9):4344-53.
9. Bruun SW, Kohler A, Adt I, Sockalingum GD, Manfait M, Martens H. Correcting attenuated total reflection-Fourier transform infrared spectra for water vapor and carbon dioxide. Appl Spectrosc. 2006 Sep;60(9):1029-39.
10. Chang, J.I.; Huang, Y.B.; Wu, P.C.; Chen, C.C.; Huang, S.C. & Tsai, Y.H. (2003) Characterization of human cervical precancerous tissue through the fourier transform infrared microscopy with mapping method. Gynecol Oncol. 2003 Dec;91(3):577-83.
11. Chekhun, V.F.; Solyanik, G.I.; Kulik, G.I.; Tryndiak, V.P.; Todor, I.N.; Dovbeshko, G.I.; & Repnytska, O.P. (2002) The SEIRA spectroscopy data of nucleic acids and phospholipids from sensitive- and drug-resistant rat tumours. J Exp Clin Cancer Res. 2002 Dec;21(4):599-607.
12. Chiriboga, L.; Xie, P.; Yee, H.; Vigorita, V.; Zarou, D.; Zakim, D. & Diem, M. (1998a) Infrared spectroscopy of human tissue. I. Differentiation and maturation of epithelial cells in the human cervix. Biospectroscopy. 1998;4(1):47-53.
13. Chiriboga, L.; Xie, P.; Vigorita, V.; Zarou, D.; Zakim, D. & Diem, M. (1998b) Infrared spectroscopy of human tissue. II. A comparative study of spectra of biopsies of cervical squamous epithelium and of exfoliated cervical cells. Biospectroscopy. 1998;4(1):55-9.
14. Chiriboga, L.; Xie, P.; Yee, H.; Zarou, D.; Zakim, D. & Diem, M..(1998c).Infrared spectroscopy of human tissue. IV. Detection of dysplastic and neoplastic changes of human cervical tissue via infrared microscopy. Cell Mol Biol.. 1998 Feb;44(1):219-29.
15. Choo-Smith, L.P.; Maquelin, K.; van Vreeswijk, T.; Bruining, H.A.; Puppels, G.J.; Ngo Thi, N.A.; Kirschner, C.; Naumann, D.; Ami, D.; Villa, A.M.; Orsini, F.; Doglia, S.M.; Lamfarraj, H.; Sockalingum, G.D.; Manfait, M.; Allouch, P. & Endtz, H.P. (2001) Investigating microbial (micro)colony heterogeneity by vibrational spectroscopy. Appl Environ Microbiol. 2001 Apr;67(4):1461-9.

16. Cohenford, M.A.; Godwin, T.A.; Cahn, F.; Bhandare, P.; Caputo, T.A. & Rigas, B. (1997) Infrared spectroscopy of normal and abnormal cervical smears: evaluation by principal component analysis. Gynecol Oncol. 1997 Jul;66(1):59-65.
17. Diem, M.; Chiriboga, L. & Yee, H. (2000) Infrared spectroscopy of human cells and tissue. VIII. Strategies for analysis of infrared tissue mapping data and applications to liver tissue. Biopolymers. 2000;57(5):282-90.
18. Essendoubi, M.; Toubas, D.; Bouzaggou, M.; Pinon, J.M.; Manfait, M.& Sockalingum, G .D. (2005) Rapid identification of Candida species by FT-IR microspectroscopy. Biochim Biophys Acta. 2005 Aug 5;1724(3):239-47.
19. Fujioka, N.; Morimoto, Y.; Arai, T. & Kikuchi, M. (2004). Discrimination between normal and malignant human gastric tissues by Fourier transform infrared spectroscopy. Cancer Detect Prev. 2004;28(1):32-6.
20. Fukuyama, Y.; Yoshida, S.; Yanagisawa, S. & Shimizu, M. (1999) A study on the differences between oral squamous cell carcinomas and normal oral mucosas measured by Fourier transform infrared spectroscopy. Biospectroscopy. 1999;5(2):117-26.
21. Gao, T.; Feng, J. & Ci Y. (1999) Human breast carcinomal tissues display distinctive FTIR spectra: implication for the histological characterization of carcinomas. Anal Cell Pathol. 1999;18(2):87-93.
22. Goodacre, R.; Timmins, E.M. Rooney, P.J.; Rowland, J.J. & Kell, D.B. (1996). Rapid identification of Streptococcus and Enterococcus species using diffuse reflectanceabsorbance Fourier transform infrared spectroscopy and artificial neural networks. FEMS Microbiol Lett. 1996 Jul 1;140(2-3):233-9.
23. Hammody, Z.; Sahu, R.K.; Mordechai, S.; Cagnano, E. & Argov, S. (2005). Characterization of malignant melanoma using vibrational spectroscopy. Scientific World Journal. 2005; 5:173-82.
24. Hammody, Z.; Huleihel, M.; Salman, A.; Argov, S.; Moreh, R.; Katzir, A. & Mordechai, S. (2007). Potential of 'flat' fibre evanescent wave spectroscopy to discriminate between normal and malignant cells in vitro. J Microsc. 2007 Nov;228(Pt 2):200-10.
25. Hammody, Z.; Argov, S.; Sahu, R.K.; Cagnano, E.; Moreh, R. & Mordechai S. (2008). Distinction of malignant melanoma and epidermis using IR micro-spectroscopy and statistical methods. Analyst. 2008 Mar;133(3):372-8.
26. Harvey, T.J.; Henderson, A.; Gazi, E.; Clarke, N.W.; Brown, M.; Faria, E.C.; Snook, R.D. & Gardner, P. (2007). Discrimination of prostate cancer cells by reflection mode FTIR photoacoustic spectroscopy. Analyst. 2007 Apr;132(4):292-5.

REFERENCES

27. Katukuri, V.K.; Hargrove, J.; Miller, S.J.; Rahal, K.; Kao, J.Y.; Wolters, R.; Zimmermann, E.M.; Wang, T.D. (2010). Detection of colonic inflammation with Fourier transform infrared spectroscopy using a flexible silver halide fiber. Biomed Opt Express. 2010 Sep 21;1(3):1014-1025.
28. Khanmohammadi, M.; Nasiri, R.; Ghasemi, K.; Samani, S.; & Bagheri Garmarudi A. (2007). Diagnosis of basal cell carcinoma by infrared spectroscopy of whole blood samples applying soft independent modeling class analogy. J Cancer Res Clin Oncol. 2007 Dec;133(12):1001-10.
29. Kirschner, C.; Maquelin, K.; Pina, P.; Ngo Thi, N.A.; Choo-Smith, L.P.; Sockalingum, G.D.; Sandt, C.; Ami, D.; Orsini, F.; Doglia, S.M.; Allouch, P.; Mainfait, M.; Puppels, G.J. & Naumann D. (2001) Classification and identification of enterococci: a comparative phenotypic, genotypic, and vibrational spectroscopic study. J Clin Microbiol. 2001 May;39(5):1763-70.
30. Krafft, C.; Shapoval, L.; Sobottka, S.B.; Schackert, G. & Salzer, R. (2006). Identification of primary tumors of brain metastases by infrared spectroscopic imaging and linear discriminant analysis. Technol Cancer Res Treat. 2006 Jun;5(3):291-8.
31. Krasteva, M.& Barth, A. (2007). Structures of the Ca2+-ATPase complexes with ATP, AMPPCP and AMPPNP. An FTIR study. Biochim Biophys Acta. 2007 Jan;1767(1):114- 23.
32. Krishna, C.M; Kegelaer, G.; Adt, I.; Rubin, S.; Kartha, V.B.; Manfait, M. & Sockalingum, G.D. (2005) Characterisation of uterine sarcoma cell lines exhibiting MDR phenotype by vibrational spectroscopy. Biochim Biophys Acta. 2005 Nov 15;1726(2):160-7.
33. Krishna, C.M.; Kegelaer, G.; Adt, I.; Rubin, S.; Kartha, V.B.; Manfait, M. & Sockalingum, G.D. (2006). Combined Fourier transform infrared and Raman spectroscopic approach for identification of multidrug resistance phenotype in cancer cell lines. Biopolymers. 2006 Aug 5;82(5):462-70.
34. Krishna, C.M.; Sockalingum, G.D.; Bhat, R.A.; Venteo, L.; Kushtagi, P.; Pluot, M. & Manfait, M. (2007). FTIR and Raman microspectroscopy of normal, benign, and malignant formalin-fixed ovarian tissues. Anal Bioanal Chem. 2007 Mar;387(5):1649-56.
35. Kohler, A.; Sulé-Suso, J.; Sockalingum, G.D.; Tobin, M.; Bahrami, F.; Yang, Y.; Pijanka, J.; Dumas, P.; Cotte, M.; van Pittius, D.G.; Parkes, G. & Martens, H.(2008). Estimating and correcting mie scattering in synchrotron-based microscopic fourier transform infrared spectra by extended multiplicative signal correction. Appl Spectrosc. 2008 Mar;62(3):259-66.
36. Kumar, S. & Barth, A.(2010). Phosphoenolpyruvate and Mg2+ binding to pyruvate kinase monitored by infrared spectroscopy. Biophys J. 2010 May 19;98(9):1931-40. Lasch, P.; Diem, M.; Hänsch, W. & Naumann, D. (2007)

Artificial neural networks as supervised techniques for FT-IR microspectroscopic imaging. J Chemom. 2007 Mar 28;20(5):209-220.
37. Lasch, P. & Naumann, D. (1998) FT-IR microspectroscopic imaging of human carcinoma thin sections based on pattern recognition techniques. Cell Mol Biol . 1998 Feb;44(1):189- 202.
38. Lee, J.; Gazi, E.; Dwyer, J.; Brown, M.D.; Clarke, N.W.; Nicholson, J.M. & Gardner, P.(2007). Optical artefacts in transflection mode FTIR microspectroscopic images of single cells on a biological support: the effect of back-scattering into collection optics. Analyst. 2007 Aug;132(8):750-5.
39. Li, Q.B.; Xu, Z.; Zhang, N.W.; Zhang, L.; Wang, F.; Yang, L.M.; Wang, J.S.; Zhou, S.; Zhang, Y.F.; Zhou, X.S.; Shi, J.S. & Wu, J.G.(2005). In vivo and in situ detection of colorectal cancer using Fourier transform infrared spectroscopy. World J Gastroenterol. 2005 Jan 21;11(3):327-30.
40. Liquier, J., Taillandier, E. . In Infrared spectroscopy of biomolecules; H.H. Mantsch, D. C., Ed.; Wiley-Liss, John Wiley & Sons, INC., Publication: NY, 1996, p 131-158.
41. Lucas, P.; Le Coq, D.; Juncker, C.; Collier, J.; Boesewetter, D.E.; Boussard-Plédel, C.; Bureau, B. & Riley, M.R. (2005). Evaluation of toxic agent effects on lung cells by fiber evanescent wave spectroscopy. Appl Spectrosc. 2005 Jan;59(1):1-9.
42. Ly, E.; Piot, O.; Wolthuis, R.; Durlach, A.; Bernard, P. & Manfait M. (2008). Combination of FTIR spectral imaging and chemometrics for tumour detection from paraffinembedded biopsies. Analyst. 2008 Feb;133(2):197-205.
43. Lyman, D.J. & Murray-Wijelath, J. (2005). Fourier transform infrared attenuated total reflection analysis of human hair: comparison of hair from breast cancer patients with hair from healthy subjects. Appl Spectrosc. 2005 Jan;59(1):26-32.
44. Mackanos, M.A. & Contag, C.H. (2010). Fiber-optic probes enable cancer detection with FTIR spectroscopy. Trends Biotechnol. 2010 Jun;28(6):317-23. Malins, D.C.; Polissar, N.L & Gunselman, S.J.(1997). Infrared spectral models demonstrate that exposure to environmental chemicals leads to new forms of DNA. Proc Natl Acad Sci . 1997 Apr 15; 94(8):3611-5.
45. Malins, D.C.; Johnson, P.M.; Barker, E.A.; Polissar, N.L.; Wheeler, T.M. & Anderson K.M. (2003). Cancer-related changes in prostate DNA as men age and early identification of metastasis in primary prostate tumors. Proc Natl Acad Sci. 2003 Apr 29;100(9):5401-6.
46. Malins, D.C.; Gilman, N.K.; Green, V.M.; Wheeler, T.M.; Barker, E.A. & Anderson, K.M. (2005). A cancer DNA phenotype in healthy prostates, conserved in tumors and adjacent normal cells, implies a relationship to carcinogenesis. Proc Natl Acad Sci 2005 Dec 27;102(52):19093-6.

REFERENCES

47. Malik, Z.; Dishi, M & Garini Y.(1996) Fourier transform multipixel spectroscopy and spectral imaging of protoporphyrin in single melanoma cells. Photochem Photobiol. 1996 May; 63(5):608-14.
48. Maquelin, K.; Kirschner, C.; Choo-Smith, L.P.; Ngo-Thi, N.A.; van Vreeswijk, T.; Stämmler, M.; Endtz, H.P.; Bruining, H.A.; Naumann, D. & Puppels, G.J. (2003). Prospective study of the performance of vibrational spectroscopies for rapid identification of bacterial and fungal pathogens recovered from blood cultures. J Clin Microbiol. 2003 Jan;41(1):324-9.
49. Mark, S.; Sahu, R.K.; Kantarovich, K.; Podshyvalov, A.; Guterman, H.; Goldstein, J.; Jagannathan, R.; Argov, S. & Mordechai, S.(2004) Fourier transform infrared microspectroscopy as a quantitative diagnostic tool for assignment of premalignancy grading in cervical neoplasia. J Biomed Opt. 2004 May-Jun;9(3):558- 67.
50. Maziak, D.E.; Do, M.T.; Shamji, F.M.; Sundaresan, S.R.; Perkins, D.G. & Wong, P.T. (2007). Fourier-transform infrared spectroscopic study of characteristic molecular structure in cancer cells of esophagus: an exploratory study. Cancer Detect Prev. 2007;31(3):244-53.
51. Mordechai, S.; Sahu, R.K.; Hammody, Z.; Mark, S.; Kantarovich, K.; Guterman, H.; Podshyvalov, A.; Goldstein, J. & Argov S. (2004) Possible common biomarkers from FTIR microspectroscopy of cervical cancer and melanoma. J Microsc. 2004 Jul;215(Pt 1):86-91.
52. Mourant, J.R.; Gibson, R.R.; Johnson, T.M.; Carpenter, S.; Short, K.W.; Yamada, Y.R.; & Freyer, J.P.(2003b). Methods for measuring the infrared spectra of biological cells. Phys Med Biol. 2003 Jan 21;48(2):243-57.
53. Mourant, J.R.; Short, K.W.; Carpenter, S.; Kunapareddy, N.; Coburn, L.; Powers, T.M. & Freyer, J.P. (2005). Biochemical differences in tumorigenic and nontumorigenic cells measured by Raman and infrared spectroscopy. J Biomed Opt. 2005 MayJun;10(3):031106.
54. Oust, A.; Møretrø, T.; Naterstad, K.; Sockalingum, G.D.; Adt, I.; Manfait, M. & Kohler A. (2006). Fourier transform infrared and raman spectroscopy for characterization of Listeria monocytogenes strains. Appl Environ Microbiol. 2006 Jan;72(1):228-32.
55. Parker, F.S. (1971). Application of Infrared spectroscopy in Biochemistry, Biology and Medicine. Plenum. NY.
56. Podshyvalov, A.; Sahu, R.K.; Mark, S.; Kantarovich, K.; Guterman, H.; Goldstein, J.; Jagannathan, R.; Argov, S. & Mordechai, S. (2005). Distinction of cervical cancer biopsies by use of infrared microspectroscopy and probabilistic neural networks. Appl Opt. 2005 Jun 20;44(18):3725-34.
57. Prathiba, J. & Malathi, R. (2008). Probing RNA-antibiotic interactions: a FTIR study. Mol Biol Rep. 2008 Mar;35(1):51-7.

58. Ramesh, J.; Salman, A.; Hammody, Z.; Cohen, B.; Gopas, J.; Grossman, N. & Mordechai, S. (2001). FTIR microscopic studies on normal and H-ras oncogene transfected cultured mouse fibroblasts. Eur Biophys J. 2001 Aug;30(4):250-5.
59. Rigas, B.; LaGuardia, K.; Qiao, L.; Bhandare, P.S.; Caputo, T & Cohenford, M.A. (2000) Infrared spectroscopic study of cervical smears in patients with HIV: implications for cervical carcinogenesis. J Lab Clin Med. 2000 Jan; 135(1):26-31.
60. Rigas, B.; Morgello, S.; Goldman, I.S. & Wong, P.T.(1990) Human colorectal cancers display abnormal Fourier-transform infrared spectra. Proc Natl Acad Sci. 1990 Oct;87(20):8140-4
61. Romeo, M.J.; Wood, B.R.; Quinn, M.A. & McNaughton, D. (2003). Removal of blood components from cervical smears: implications for cancer diagnosis using FTIR spectroscopy. Biopolymers. 2003;72(1):69-76.
62. Sahu, R.K.; Argov, S.; Bernshtain, E.; Salman, A.; Walfisch, S.; Goldstein, J. & Mordechai, S. (2004a). Detection of abnormal proliferation in histologically 'normal' colonic biopsies using FTIR-microspectroscopy. Scand J Gastroenterol. 2004 Jun;39(6):557-66.
63. Sahu, R.K.; Argov, S.; Salman, A.; Huleihel, M.; Grossman, N.; Hammody, Z.; Kapelushnik, J. & Mordechai, S. (2004b). Characteristic absorbance of nucleic acids in the Mid-IR region as possible common biomarkers for diagnosis of malignancy. Technol Cancer Res Treat.2004 Dec;3(6):629-38.
64. Sahu, R.K.; Argov, S.; Salman, A.; Zelig, U.; Huleihel, M.; Grossman, N.; Gopas, J.; Kapelushnik, J. & Mordechai, S. (2005). Can Fourier transform infrared spectroscopy at higher wavenumbers (mid IR) shed light on biomarkers for carcinogenesis in tissues? J Biomed Opt. 2005 Sep-Oct;10(5):054017.
65. Sahu, R.K.; Zelig, U.; Huleihel, M.; Brosh, N.; Talyshinsky, M.; Ben-Harosh, M.; ,Mordechai, S. & Kapelushnik, J. (2006a) Continuous monitoring of WBC (biochemistry) in an adult leukemia patient using advanced FTIR-spectroscopy. Leuk Res. 2006; 30(6):687-93.
66. Sahu, R.K.; Mordechai, S.; Pesakhov, S.; Dagan, R. & Porat, N. (2006b) Use of FTIR Spectroscopy to Distinguish between Capsular Types and Capsular Quantities in Streptococcus pneumoniae. Biopolymers. 2006; 83(4):434-442.
67. Sahu, R. K.; Mordechai,S. & Manor, E. (2008) Nucleic acids absorbance in Mid IR and its effect on diagnostic variates during cell division: a case study with lymphoblastic cells. Biopolymers. 2008 Nov;89(11):993-1001.
68. Sahu, R.K.; Argov, S.; Walfisch, S.; Bogomolny, E.; Moreh, R. & Mordechai, S. (2010). Prediction potential of IR-micro spectroscopy for colon cancer relapse. Analyst. 2010 Mar;135(3):538-44.

REFERENCES

69. Salman, A.; Ramesh, J.; Erukhimovitch, V.; Talyshinsky, M.; Mordechai, S. & Huleihel, M. (2003). FTIR microspectroscopy of malignant fibroblasts transformed by mouse sarcoma virus. J Biochem Biophys Methods. 2003 Feb 28;55(2):141-53.
70. Salman, A.; Sahu, R.K.; Bernshtain,E.; Zelig, U.; Goldstein, J.; Walfisch,S.;, Arov, S. & Mordechai,S. (2004). Probing cell proliferation in the human colon using vibrational spectroscopy: a novel use of FTIR-Microspectroscopy. Vib. spec. 2004. 34:301-308.
71. Salman, A.; Pomerantz, A.; Tsror, L.; Lapidot, I.; Zwielly, A.; Moreh, R.; Mordechai, S. & Huleihel, M. (2011). Distinction of Fusarium oxysporum fungal isolates (strains) using FTIR-ATR spectroscopy and advanced statistical methods. Analyst. 2011 Mar 7;136(5):988-95.
72. Sandt, C.; Sockalingum, G.D.; Aubert, D.; Lepan, H.; Lepouse, C.; Jaussaud, M.; Leon, A.; Pinon, J.M.; Manfait, M. & Toubas, D. (2003) Use of Fourier-transform infrared spectroscopy for typing of Candida albicans strains isolated in intensive care units. J Clin Microbiol. 2003 Mar;41(3):954-9.
73. Sandt, C.; Madoulet, C.; Kohler, A.; Allouch, P.; De Champs, C.; Manfait, M. & Sockalingum, G.D. (2006). FT-IR microspectroscopy for early identification of some clinically relevant pathogens. J Appl Microbiol. 2006 Oct;101(4):785-97.
74. Schultz, C.P.; Liu, K.; Johnston, J.B. & Mantsch, H .H. (1996) Study of chronic lymphocytic leukemia cells by FT-IR spectroscopy and cluster analysis. Leuk Res. 1996 Aug;20(8):649-55.
75. Sukuta, S. & Bruch, R. (1999) Factor analysis of cancer Fourier transform infrared evanescent wave fiberoptical (FTIR-FEW) spectra. Lasers Surg Med. 1999;24(5):382-8.
76. Toubas, D.; Essendoubi, M.; Adt, I.; Pinon, J.M.; Manfait, M. & Sockalingum, G.D. (2007). FTIR spectroscopy in medical mycology: applications to the differentiation and typing of Candida. Anal Bioanal Chem. 2007 Mar;387(5):1729-37.
77. Wang, H.P.; Wang, H.C. & Huang YJ. (1997) Microscopic FTIR studies of lung cancer cells in pleural fluid. Sci Total Environ. 1997 Oct 1; 204(3):283-7.
78. Wong, P.T.; Capes, S.E. & Mantsch, H.H.; Hydrogen bonding between anhydrous cholesterol and phosphatidylcholines: an infrared spectroscopic study. Biochim Biophys Acta. 1989 Mar 27;980(1):37-41.
79. Wood, B.R.; Quinn, M.A.; Tait, B.; Ashdown, M.; Hislop, T.; Romeo, M. & McNaughton, D. (1998). FTIR microspectroscopic study of cell types and potential confounding variables in screening for cervical malignancies. Biospectroscopy. 1998;4(2):75-91.

80. Wu, J.G.; Xu, Y.Z.; Sun, C.W.; Soloway, R.D.; Xu, D.F.; Wu, Q.G.; Sun, K.H.; Weng, S.F. & Xu, G.X.. Distinguishing malignant from normal oral tissues using FTIR fiber-optic techniques. Biopolymers. 2001;62(4):185-92.
81. Yang, D.; Castro, D.J.; el-Sayed, I.H.; el-Sayed, M.A.; Saxton, R.E & Zhang NY.(1995) A Fourier-transform infrared spectroscopic comparison of cultured human fibroblast and fibrosarcoma cells: a new method for detection of malignancies. J Clin Laser Med Surg. 1995 Apr;13(2):55-
82. Yano, K.; Ohoshima, S.; Gotou, Y.; Kumaido, K.; Moriguchi, T. & Katayama H. (2000) Direct measurement of human lung cancerous and noncancerous tissues by Fourier transform infrared microscopy: can an infrared microscope be used as a clinical tool? Anal Biochem. 2000 Dec 15;287(2):218-25.
83. Zelig, U.; Kapelushnik, J.; Moreh, R.; Mordechai, S. & Nathan, I. (2009). Diagnosis of cell death by means of infrared spectroscopy. Biophys J. 2009 Oct 7;97(7):2107-14.
84. Zelig, U.; Mordechai, S.; Shubinsky, G.; Sahu, R.K.; Huleihel, M.; Leibovitz, E.; Nathan, I. & Kapelushnik, J.(2011). Pre-screening and follow-up of childhood acute leukemia using biochemical infrared analysis of peripheral blood mononuclear cells. Biochim Biophys Acta. 2011 Sep;1810(9):827-35.
85. Zhang, L.; Small, G.W.; Haka, A.S.; Kidder, L.H. & Lewis, E.N. (2003) Classification of Fourier transform infrared microscopic imaging data of human breast cells by cluster analysis and artificial neural networks. Appl Spectrosc. 2003 Jan;57(1):14-22.

CHAPTER 11

Hyperspectral Imaging and Chemometric Modeling of Echinacea — A Novel Approach in the Quality Control of Herbal Medicines

Maxleene Sandasi 1, Ilze Vermaak 1, Weiyang Chen 1 and Alvaro M. Viljoen 1,2,*

[1]Department of Pharmaceutical Sciences, Tshwane University of Technology, Private Bag X680, Pretoria 0001, South Africa
[2]Department of Pharmaceutics and Industrial Pharmacy, Faculty of Pharmacy King Abdulaziz University, Jeddah 21589, Saudi Arabia

ABSTRACT

Echinacea species are popularly included in various formulations to treat upper respiratory tract infections. These products are of commercial importance, with a collective sales figure of $132 million in 2009. Due to their close taxonomic alliance it is difficult to distinguish between the three *Echinacea* species and incidences of incorrectly labeled commercial products have been reported. The potential of hyperspectral imaging as a rapid quality control method for raw material and products containing *Echinacea* species was investigated. Hyperspectral images of root and leaf material of authentic *Echinacea* species (*E. angustifolia*, *E. pallida* and *E. purpurea*) were acquired using a sisuChema shortwave infrared (SWIR) hyperspectral pushbroom imaging system with a spectral range of 920–2514 nm. Principal component analysis (PCA) plots showed a clear distinction between the root and leaf samples of the three *Echinacea* species and further differentiated the roots of different species. A classification model with a high coefficient of determination was constructed to predict the identity of the species included in commercial products. The majority of products (12 out of 20) were convincingly predicted as containing *E. purpurea*, *E. angustifolia* or both. The use of ultra performance liquid chromatography-mass spectrometry (UPLC-MS) in the differentiation of the species presented a challenge due to chemical similarities between the solvent extracts. The results show that hyperspectral imaging is an objective and non-destructive quality control method for authenticating raw material.

Keywords: *Echinacea*; chemometrics; hyperspectral imaging; principal component analysis; partial least squares discriminant analysis; quality control

1. INTRODUCTION

Echinacea species have a long history of use as traditional medicine to treat various ailments, including infections such as syphilis and septic wounds, cough, sore throat and tonsillitis. The three species currently used medicinally and most commonly to treat the common cold, influenza-like illnesses and upper respiratory tract infections (URTIs) are *Echinacea angustifolia* DC., *Echinacea pallida* (Nutt.) Nutt. and *Echinacea purpurea* (L.) Moench [1,2]. *Echinacea* species (Figure 1) have collectively been referred to as purple cornflower and pre-1968, *E. angustifolia* and *E. pallida* were considered to be varieties of the same species. *Echinacea pallida* is sometimes referred to as pale coneflower or pale purple coneflower and *E. angustifolia* is known as narrow leaf purple coneflower or Kansas snakeroot. A revision of the species suggested the following nomenclature;*E. pallida* var. *angustifolia* (DC.) Cronq. and *E. pallida* var. *pallida* (Nutt.) Cronq but the three species (*E. angustifolia*, *E. pallida* and *E. purpurea*) are still separately indicated on commercial products and will therefore be referred to as such in this study. Clinical trials to investigate the effectiveness of*Echinacea* preparations in the treatment of URTIs have reported effects superior to placebo, but the phytochemical diversity makes the interpretation of research findings difficult [1]. A structured review on clinical trials concluded that *Echinacea* use has no benefit in the treatment and prevention of colds [2]. Despite the publication of some negative research results, *Echinacea*products remain extremely popular, and are still frequently mentioned and/or promoted in the media and are commercially important. In the United States of America, annual sales of all*Echinacea* products exceeded $200 million in 2000 and 2001 whereafter it steadily declined to about $129 million in 2006 [3]. According to the Nutrition Business Journal, *Echinacea* sales rose to $132 million in 2009 [4] and it is still one of the best-selling herbal preparations in the USA [1].

There is variation in the chemical constituents of *Echinacea* between the different species and within different plant parts. The main chemical constituents include alkamides, phenylpropanoids, polysaccharides and volatile oils, as well as minor constituents such as flavonoids. The chemical composition, including alkamide content, determined using liquid chromatography, is traditionally used to establish the quality of plant material and preparations, as well as to identify the species [1]. According to a review by Barnes [1], several poor quality products have been identified, some not stating the species or plant parts used and others purporting to contain a certain species while in fact it does not. Of 59 commercial products analysed, 28 (48%) did not contain the species claimed on the label. Clearly, developing methods that can rapidly authenticate plant material will have a positive impact on the quality of products distributed to consumers.

1. INTRODUCTION

Figure 1. Photographs of the three *Echinacea* species: (**a**) *E. angustifolia*; (**b**) *E. pallida*; and (**c**) *E. purpurea*.

While the interchangeable use of the three *Echinacea* species in commercial products has not been reported as posing any potential risk, evidence of pharmacological equivalence of these species has not been clearly demonstrated. The phytochemical diversity in the raw materials (species) and commercial products that are widely available as directly pressed juices, ethanolic or hydrophilic extracts, powdered leaves, flowers and roots as well as extracts standardised to marker constituents such as phenolic acid or echinacoside content may raise concerns relating to efficacy [5]. Research on the efficacy of *Echinacea* raw materials and commercial products as immunomodulating agents has been conducted both *in vitro* and *in vivo*. Evidence of varying degrees of efficacy of *Echinacea* was demonstrated in mouse models where *E. purpurea* demonstrated superior immunomodulation activity compared to *E. angustifolia* and *E. pallida* [6]. In a separate study, *E. purpurea* whole powder again demonstrated consistently high macrophage activation, while products standardised to marker constituents were found to be inactive as immunomodulators [7]. A high degree of variability among similarly standardised extracts was also observed. A comprehensive review on the quality and pharmacological activities of *Echinacea* raw materials and commercial products reports on the inconsistencies in product quality while pre-clinical and clinical studies on many commercial products do not demonstrate pharmacological equivalence among products [3]. This variation between products and subsequent implications on efficacy should be considered important consumer information which may assist in the choice of product. To date however, there is no consensus with regards to the species, plant part or extraction methods that yields a product with the most desirable pharmacological activity.

Hyperspectral imaging (HSI) acquires both spectral and spatial information from a sample through a combination of conventional spectroscopy and imaging [8]. It has been used with success as an analytical tool to assess raw material and product quality in the agricultural, food and beverage and pharmaceutical industries amongst others [8,9], where non-destructive analyses can be performed in a much shorter time compared to conventional analysis methods such as liquid chromatography. In hyperspectral imaging, a hypercube is created through the combination of two spatial (x;y) and one wavelength (λ) dimension (Figure 2). Images are collected as a function of wavelength resulting in a stack of images referred to as a hypercube. At any wavelength, the image comprises of several pixels (2D-squares) where each pixel represents a spectrum containing point chemical information. The images are then analysed to identify wave regions where chemical differences are observed within the sample under investigation (Figure 2). The inclusion of spatial information yields more information about a sample and unknown samples can be more accurately predicted in some cases as compared to single-point spectroscopy. Multivariate analysis tools such as principal component analysis (PCA) and partial least squares (PLS) analysis are used to reduce the high dimensionality of data and to display compositional differences in an image enabling the development of models that can be used for quality control purposes [8,10,11].

The aim of this study was to investigate the potential of hyperspectral imaging in combination with multivariate data analysis to distinguish between the three commercially important *Echinacea* species and to correctly identify the *Echinacea* species present in commercial products.

2. RESULTS AND DISCUSSION

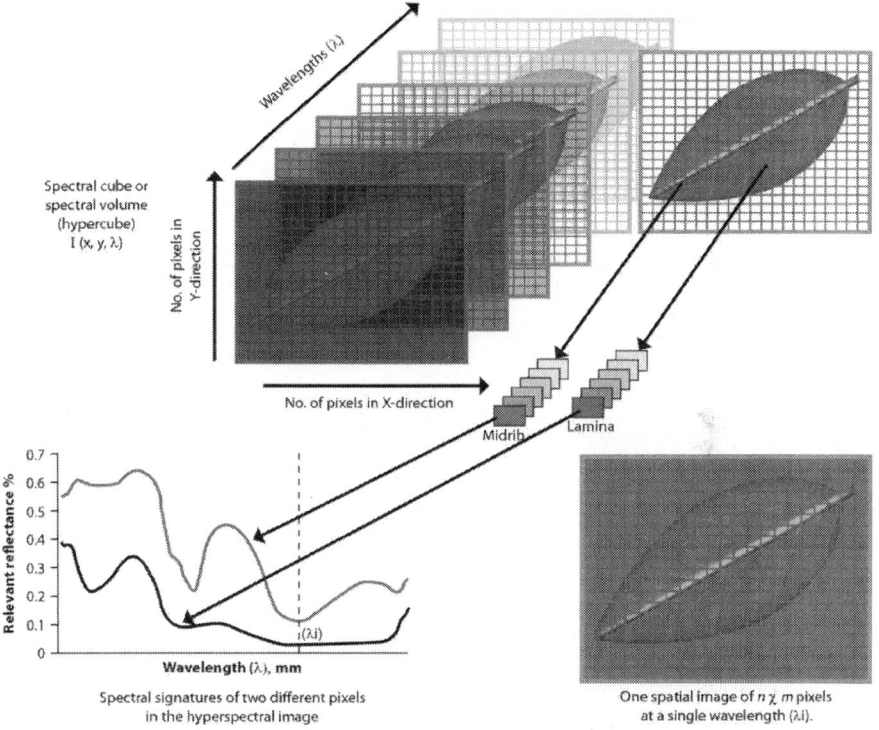

Figure 2. An illustration of a spectral hypercube created through the combination of two spatial (x;y) and one wavelength (λ) dimension in hyperspectral imaging.

2. RESULTS AND DISCUSSION

2.1. Image Analysis Using PCA

Figure 3 displays NIR spectra of both leaf (a) and root (b) powders for the three *Echinacea* species. The spectral patterns do not show distinctive features that can be used to differentiate the species and hence the need for PCA which provides visual plots such as score images and scatter plots to observe clearer differences.

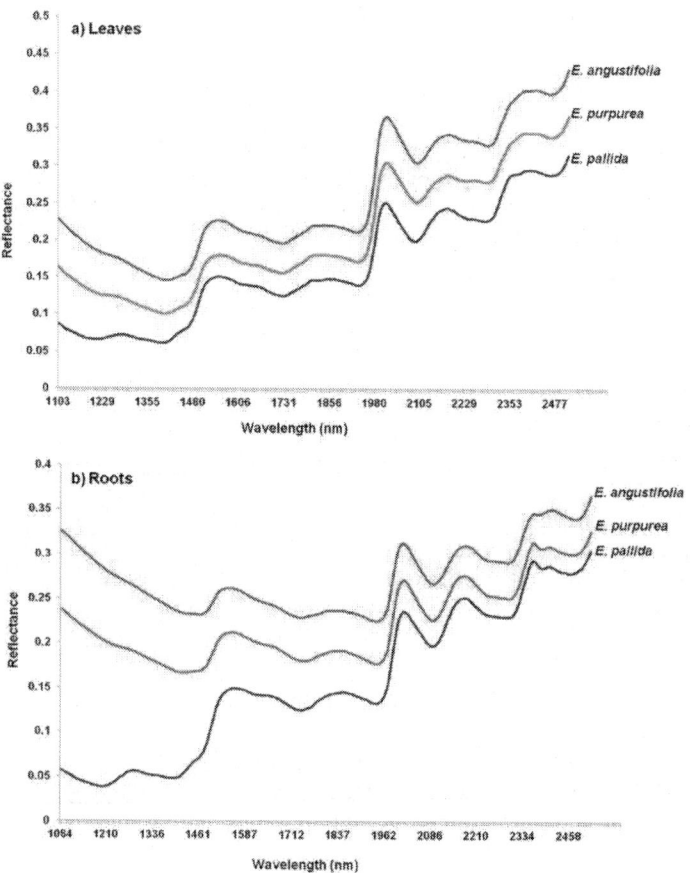

Figure 3. NIR imaging spectra of (**a**) *Echinacea* leaves and (**b**) roots showing similar spectral patterns of the three species.

The interactive score image and scatter plots were coloured according to the score values of PCs. The score image is an amplitude plot where similar colours represent similar score values and *vice-versa*. Principal component analysis of the root and leaf image showed a clear separation of *E. pallida* from the other two species based on the colour amplitude, as observed in the PCA score image of the first PC (t_1) after mean centering and standard normal variate (SNV) correction was applied to the data (Figure 4a). Replicate *E. pallida* root samples (EPaR) showed a consistently distinct high colour amplitude (yellow-red) compared to the other species, evidence that its chemical profile differed significantly from the other species. As observed for the root powders, *E. pallida* leaf (EPaL) also showed a higher colour profile (light blue) compared to leaves of *E. angustifolia* (EAL) and *E. purpurea* (EPL) that had low amplitude of -5 (dark blue) (Figure 4a).

2. RESULTS AND DISCUSSION

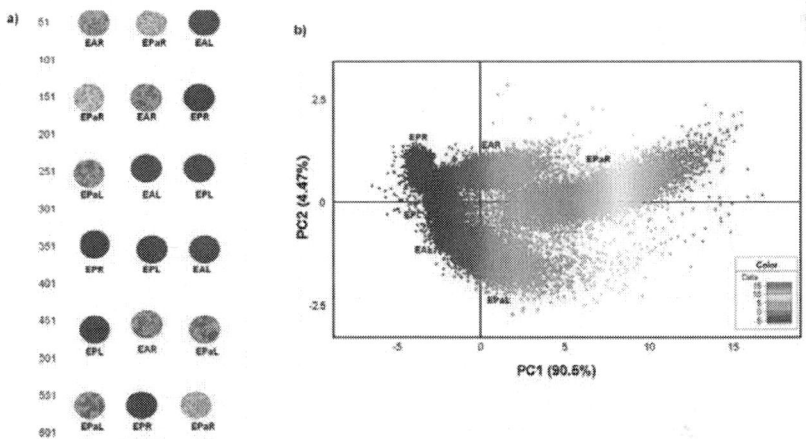

Figure 4. PCA score image (t1) of *Echinacea* root and leaf powders showing distinction of the three species based on colour amplitudes (**a**). The corresponding score plot (PC1 vs.PC2) shows distinct pixel clusters coloured according to score values that correspond to the images (**b**). (EAL—*E. angustifolia* leaf, EPL—*E. purpurea* leaf, EPaL—*E. pallida* leaf, EAR—*E. angustifolia* root, EPR—*E. purpurea* root, EPaR—*E. pallida* root).

The corresponding scatter plot shows pixels coloured according to score values which correlates to the amplitudes scale in the score image. Using PCA, chemical variation was observed within the HSI data which revealed 6 pixel clusters with each cluster representing plant parts of different species (Figure 4b). The cumulative chemical variation modeled using three principal components (PCs) was 97.2% (R^2X_{cum} = 0.972). The variation of 90.5% along PC1 was responsible for separating mainly the root powders where *E. pallida* (EPaR) was shown to be the most distinct with its pixel cluster (highest score value-red yellow) falling on the far positive PC1. The *E. angustifolia* root cluster (EAR) showed pixels with lower score values spanning Y = 0 region, while *E. purpurea* root pixels (EPR) occupied negative PC1 (Figure 4b). The observation supports results of the score image where maximum variance modeled along PC1 was shown to differentiate mainly the root samples. Distinction of the leaf samples was observed along PC2 in the scatter plot; however, the variation appears minimal (4.47%) since the presence of root samples influenced variation within the model.

To obtain a clearer picture of the chemical distinction of the root samples, it was necessary to model these separately from the leaves to eliminate the influence of the leaf chemistry on the model. Figure 5a is a score image of PC1 which showed three distinct colour amplitudes, each representing a different species. As observed in Figure 4a, *E. pallida* displayed the highest colour amplitude (orange-yellow) while *E. angustifolia* (light blue) and *E. purpurea* (dark blue) showed lower amplitudes that differed slightly. The corresponding scatter 2D density plot demonstrated a clear separation of the three pixel clusters along PC1 with 94.5% chemical variation in the data cube attributed to the distinction of the three species (Figure 5b).

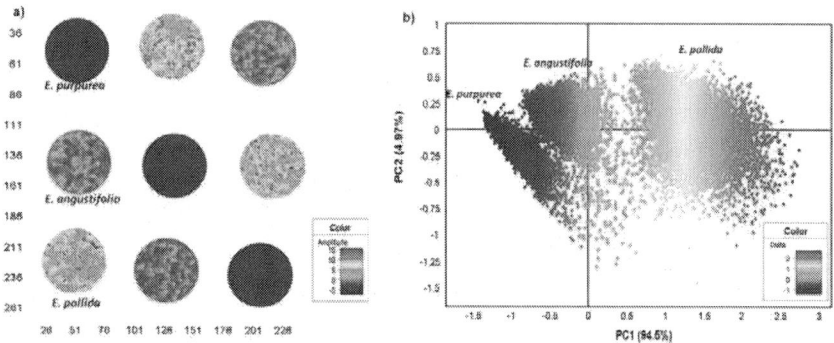

Figure 5. PCA score image (t1) of *Echinacea* root powders showing distinction of the three species based on colour amplitudes (**a**). The corresponding score plot (PC1 *vs.* PC2) shows distinct pixel clusters coloured according to score values that correspond to the images (**b**). (EAR—*E. angustifolia* root, EPR—*E. purpurea* root, EPaR—*E. pallida* root).

The distribution of the pixel clusters along PC1 was consistent with Figure 4b, only clearer. The results for the root samples demonstrate that the three *Echinacea* species have distinct chemical profiles which was easily identified using NIR chemical imaging. To analyse the leaf data clearly, the leaf samples were modeled separately to exclude the influence from the roots. Figure 6a shows the score image where *E. pallida* displayed colour distinction (light blue-yellow) while similarities between *E. angustifolia* and *E. purpurea* leaves (dark blue) were evident. The corresponding scatter 2D plot of pixels demonstrates separation of the *E. pallida* cluster from the other two species along PC1. The majority of the variance was captured in the first two PCs with PC1 accounting for 82.6% of the data and 3.94% in PC2 (Figure 6b). It is clear from the plots that the leaf chemistry of the three species is less varied as demonstrated by the variance parameters. However, *E. pallida* demonstrated chemical distinction compared to the other species as identified by HSI and multivariate data analysis techniques. To investigate the differences observed in the score images of the leaves and roots, loadings line plots of the first vector (P1) were constructed for the leaf model (Figure 7a) and the roots model (Figure 7b). The loadings plots show the region between 1937–2400 nm as carrying discriminating information of the three species based on both the leaf and the root chemistry. In both cases, positive (peaks) and negative loadings (troughs) were recorded in this region and these regions could be further investigated for molecular signals that can be assigned to specific plant species.

2. RESULTS AND DISCUSSION

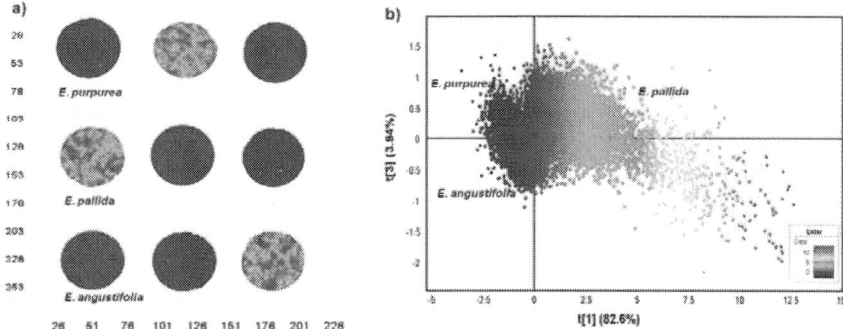

Figure 6. PCA score image (t1) of *Echinacea* leaf powders based on colour amplitudes (**a**). The corresponding score plot (PC1 *vs.* PC3) shows minimal separation of the pixel clusters (**b**). (EAL—*E. angustifolia* leaf, EPL—*E. purpurea* leaf, EPaL—*E. pallida* leaf).

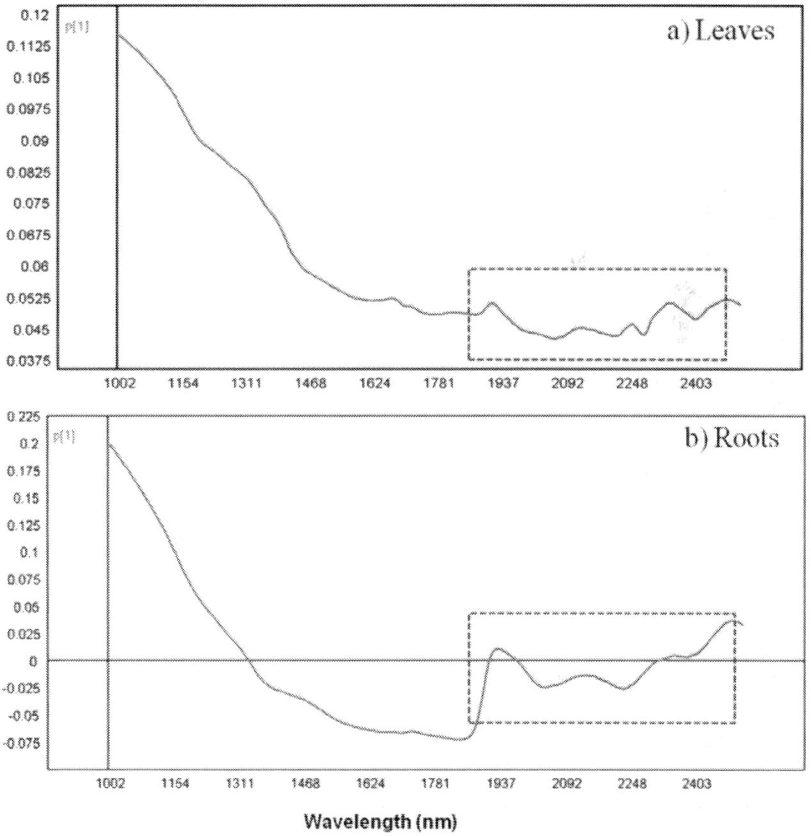

Figure 7. Loadings line plot of vector P1 for the leaf score image (**a**) and the root score image (**b**) showing variables responsible for separation of the three species.

Based on these observations, it is evident that the use of NIR chemical imaging for the qualitative differentiation of *Echinacea* species presents a promising visual technique in the quality control of the raw material. The root chemistry of the species presents a better choice for analysis; however, the minor chemical differences between leaf samples can also be detected using the imaging tool. Previous reports on the chemical profiling of *Echinacea* species have reported marked differences between *E. angustifolia* and *E. pallida* that were previously regarded as varieties of the same species until they were taxonomically revised in 1968 [12]. In this study, *E. pallida* has demonstrated a distinct chemical profile that could be observed in both root and leaf samples.

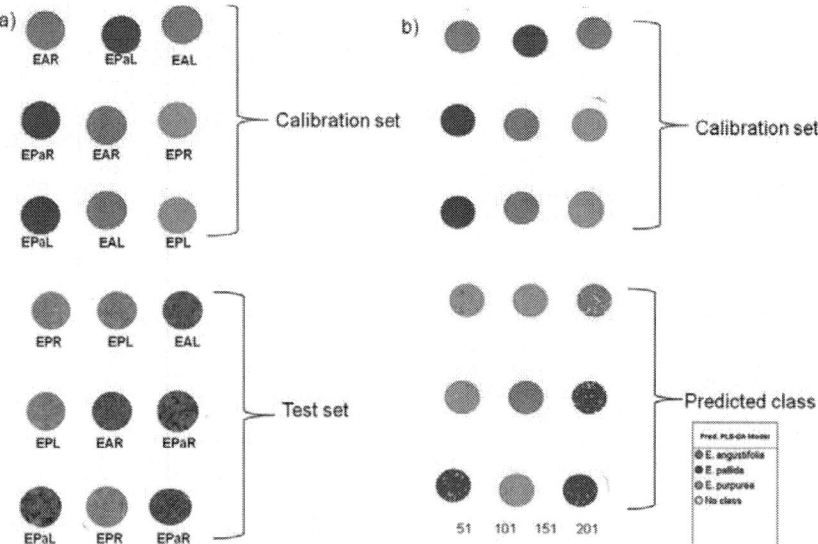

Figure 8. PLS-DA Y-image showing calibration samples used to build the model and test samples that were excluded for model validation (**a**). The predictions show test samples assigned to classes that correspond to the species imaged (**b**) (EAL—*E. angustifolia* leaf, EPL—*E. purpurea* leaf, EPaL—*E. pallida* leaf, EAR—*E. angustifolia* root, EPR—*E. purpurea* root, EPaR—*E. pallida* root).

Having successfully established that HSI can distinguish between the three *Echinacea* species, the next objective was to develop a PLS-DA model to predict the identity of commercial *Echinacea* products introduced into the model as an external dataset. Since many *Echinacea* products are prepared using both the leaf and root materials, a model that included both the leaves and the roots was constructed. Ultimately the model predictions would then identify the species present but not specify whether the material used was the root, leaf or both. Figure 8a is a PLS 1 Y-image of the PLS-DA model where the calibration set showed each species (class) assigned a different colour while the background or some unknown areas were classified as "no class" implying that the sample/region within the image could not be correlated to the modeled data. The test set samples (grey-scale) were excluded for external validation of the model where the correct identities were known. Introducing the external test set into the PLS-DA calibration model provided results shown in Figure 8b. The predicted classes of the test samples were in agreement with prior knowledge on identities of the samples based on the colour coding. There were however regions that were misclassified within the samples which can be attributed to chemical similarities between the species and hence the overlap in the pixel data. The PLS-DA model however exhibited good model statistics with R^2X_{cum} of 0.980 and cumulative variation of Y (Q^2Y_{cum}) of 0.779 that could be predicted by three components which was subsequently used for class prediction of external samples.

2.2. Class Prediction of Commercial Echinacea Samples

Twenty commercial *Echinacea* samples and four authentic *Echinacea* raw material control samples captured as a single image were introduced into the PLS-DA model for class prediction. Figure 9 shows the PLS-DA prediction image after matching the chemical profiles of the products to the authentic *Echinacea* samples.

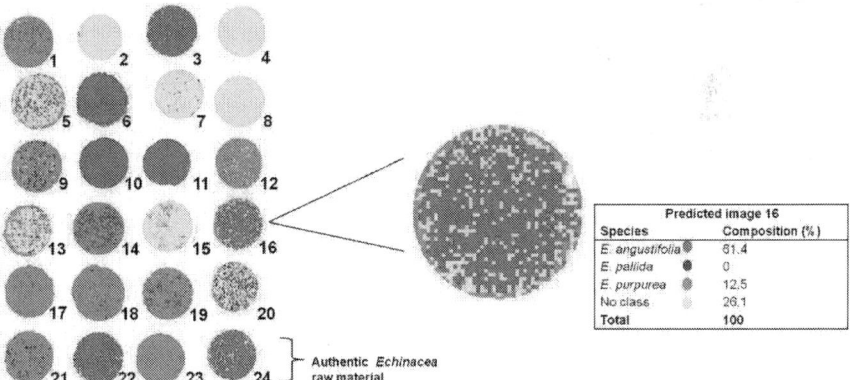

Figure 9. PLS prediction images (t_1) of *Echinacea* commercial products and authentic raw material showing variation in the chemical composition of *Echinacea* commercial products. The enlarged insert demonstrates the predicted levels of each species in product 16.

The predictions are represented by colour where *E. angustifolia* is represented in red, *E. pallida* in blue and *E. purpurea* in green. The results indicate that 12 out of 20 products were correctly classified (indicated with [a]) with the HSI class prediction matching the product label. The remaining eight products were misclassified (indicated with [b]) as the HSI prediction did not match the product label. Seven of the 20 products contained high levels of *E. purpurea* (1, 9, 12, 14, 17, 18 and 19) while five products contained high levels of *E. angustifolia* (3, 6, 10, 11 and 16). Five of the 20 products (2, 4, 7, 8, and 15) seemed to present a completely different profile to the powdered authentic *Echinacea* species and hence these were identified as no class (yellow). Upon investigation it was discovered that most of the samples predicted as 'no class' were either labeled as extracts or concentrates of *Echinacea* and not unprocessed raw material. As chemical processing alters the chemistry of the products, this presents chemical variation between the modeled and predicted samples and hence the products were not classified as *Echinacea*. In some images (5, 13 and 20) the 'no class' prediction was more prominent while trace amounts of one or two *Echinacea*species was detected. The observation can be explained by the presence of excipients such as magnesium stearate in the formulations in greater proportions compared to *Echinacea* raw material.

Multi-ingredient formulations containing other herbs such as garlic, parsley and goldenseal were also predicted as containing trace amounts of the *Echinacea* species while the majority of the constituents could not be classified (Product 20). The last sample row in the image containing authentic *Echinacea* raw material as a control demonstrates the accuracy of the model in predicting*Echinacea* raw material. The four samples (21, 22, 23 and 24) were correctly identified where the major species predicted to be dominant in the sample matched the species imaged (root and/or leaf samples). The potential of HSI in determining the presence or absence of *Echinacea* raw material in both commercial products and raw material samples has been demonstrated.

In addition to a qualitative assessment of the product composition, it was also possible to predict the percentage composition of the constituent species within a product by highlighting the species in the image and obtaining the corresponding prediction table as demonstrated in Figure 9 (insert). The quantitative predictions are reported in Table 1 together with a comparison of the product label. Table 1 shows that most of the commercial products does contain *Echinacea* raw material in high levels and most of the HSI results are in agreement with the label claim. However, HSI analysis detected the presence of more than one species in many of the products. According toTable 1, *E. purpurea* and *E. angustifolia* seem to be the raw materials of choice in many products while *E. pallida* is rarely used. Only two products were labeled as containing *E. pallida*; however, according to the predictions, *E. purpurea* was identified as occurring in higher proportions in both cases while the presence of *E. pallida* was almost negligible (<10%).

2. RESULTS AND DISCUSSION

Table 1. Results of the HSI classification analysis using the (PLS-DA) model in comparison to the product label.

Product	Product Label Claim	HSI Species Predictions (%)			
		E. angustifolia	E. purpurea	E. pallida	No Class
1	E. purpurea (herb, root) & E. angustifolia herb	1.9	89.8 [a]	0	8.3
2	E. angustifolia root extract	0	0	0.2	99.8 [b]
3	E. purpurea root & aerial parts	89.3 [a]	2.6	0.2	7.9
4	E. purpurea Herba & Radix dry concentrate	0	0	0	100 [b]
5	E. purpurea & E. angustifolia	14.3	7.7	1.2	76.8 [b]
6	E. purpurea root & E. angustifolia root	94.6 [a]	5.4	0	0.1
7	Echinacea purpurea extract	0	0	0	100 [b]
8	E. purpurea Herba & E. purpurea Radix	0	0	0	100 [b]
9	E. pallida	0	72.2 [a]	7.4	20.4
10	E. purpurea root & E. angustifolia root	86.7 [a]	12.6	0	0.7
11	E. angustifolia root	99.5 [a]	0	0.1	0.4
12	E. purpurea herb	0.2	73.2 [a]	0.2	26.4
13	E. pallida (outer label) & E. purpurea (inner label)	1.3	22.9	2.4	73.3 [b]
14	Echinacea aerial powder & root powder	9.2	90.1 [a]	0	0.7
15	E. angustifolia root & rhizome	0	0	0	100 [b]
16	E. purpurea stem, leaf, flower	61.8 [a]	12.1	0	26.1
17	E. purpurea root	0.2	95.4 [a]	0	4.4
18	Echinacea blend (angustifolia, pallida, purpurea)	0.1	92.4 [a]	2	5.6
19	Echinacea leaf powder & standardised extract	4.9	88.5 [a]	0	6.6
20	Echinacea, Goldenseal, Elderberry, Garlic & Parsley	16.3	8.1	9.4	66.3 [b]
21	E. purpurea leaf (authentic raw material)	0.1	90.2 [a]	6.5	3.3
22	E. angustifolia root (authentic raw material)	89.1 [a]	3.5	0.4	7
23	E. purpurea root (authentic raw material)	0	97.5 [a]	0	2.5
24	E. angustifolia leaf (authentic raw material)	64.2 [a]	6.4	1.8	27.5

[a] The formulation was predicted to contain the indicated species in highest proportions. [b] The formulation was predicted to contain little or no authentic Echinacea raw material.

2.3. Ultra Performance Liquid Chromatography-Mass Spectrometry (UPLC-MS) Analysis

In order to confirm the HSI predictions, UPLC-MS was used to analyse both the raw materials and the products to determine the botanical species in the products. Three solvent systems (methanol, hexane and chloroform), previously reported to provide some distinction between the three *Echinacea* species, were used. Only the methanol and hexane root extracts were used for product classification as the leaf chromatograms did not display any clear distinction between the three species. Additionally, the chloroform root profiles were very similar to that of hexane, therefore only the hexane chromatograms will be presented for discussion. The BPI chromatograms of each product were qualitatively assessed against the chromatograms of the three different species to identify similarities or differences in the chemical profiles, but marked chemical similarities of the solvent extracts of the three species were observed. Figure 10 shows the BPI chromatograms of the methanol root extracts for the authentic raw materials.

The chromatograms demonstrate marked similarities between the three species with the major peaks present in all three species.

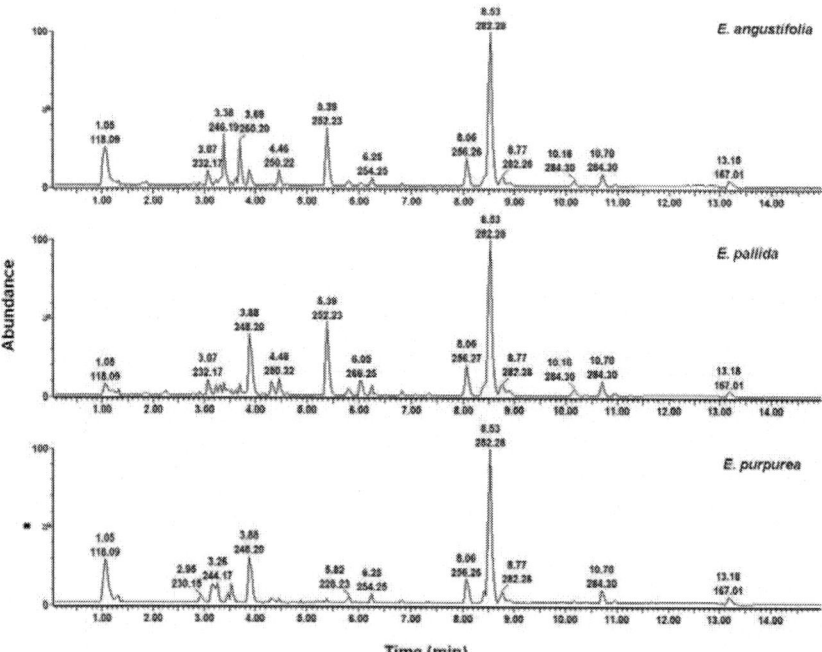

Figure 10. Chromatographic fingerprints of methanol root extracts of *Echinacea* species.

An inspection of the methanol leaf chromatograms (Figure 11) showed similarities in the major peaks, thus both methanol and hexane root extracts presented a challenge in their usefulness to differentiate between the three species. Closer inspection of the methanol chromatograms revealed that minor peaks in the region between 3–5 min appeared to be unique to a species and were therefore selected for qualitative assessment. The methanol chromatogram of product 9, which was labeled as *E. pallida* but predicted as *E. purpurea*, was compared to the chromatograms of the authentic material.

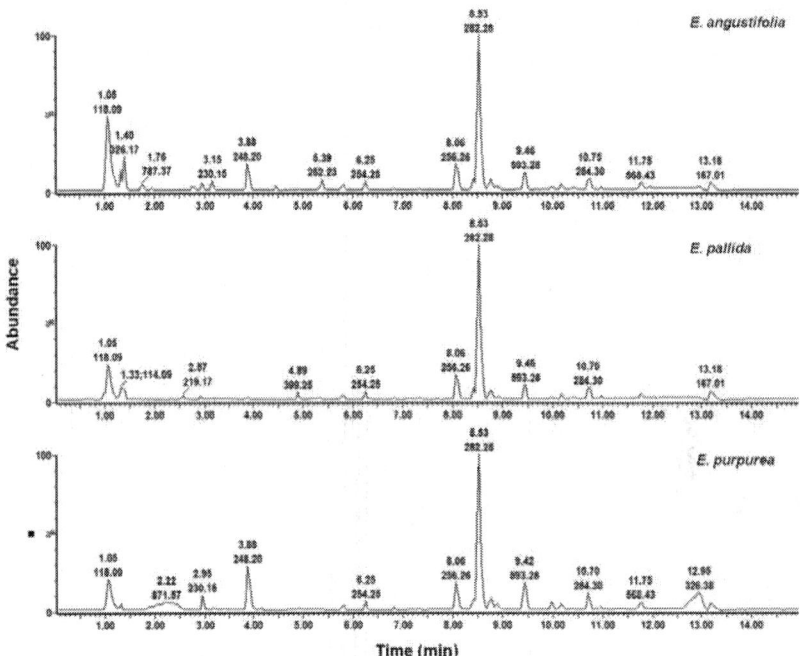

Figure 11. Chromatographic fingerprints of methanol leaf extracts of *Echinacea* species.

Figure 12 shows that product 9 was devoid of a signal present in *E. pallida* at Rt 5.39 min (mz^{-1} = 252.23). Additionally, a minor characteristic peak unique to *E. purpurea* at Rt 4.89 min (mz^{-1} = 277.22) is also present in product 9. Based on this comparison it was possible to confirm the HSI prediction that product 9 contains *E. purpurea* instead of *E. pallida* as claimed on the product label. The shortfalls of using UPLC-MS as a tool for the qualitative differentiation of *Echinacea* species and classification of commercial products were demonstrated due to the chemical similarities of the solvent extracts obtained from the raw materials. On the other hand, because HSI presents chemical profiles of the whole metabolome in an untargeted approach, it allows for any possible differences in the composition of the raw material to be detected using this approach in a simple manner.

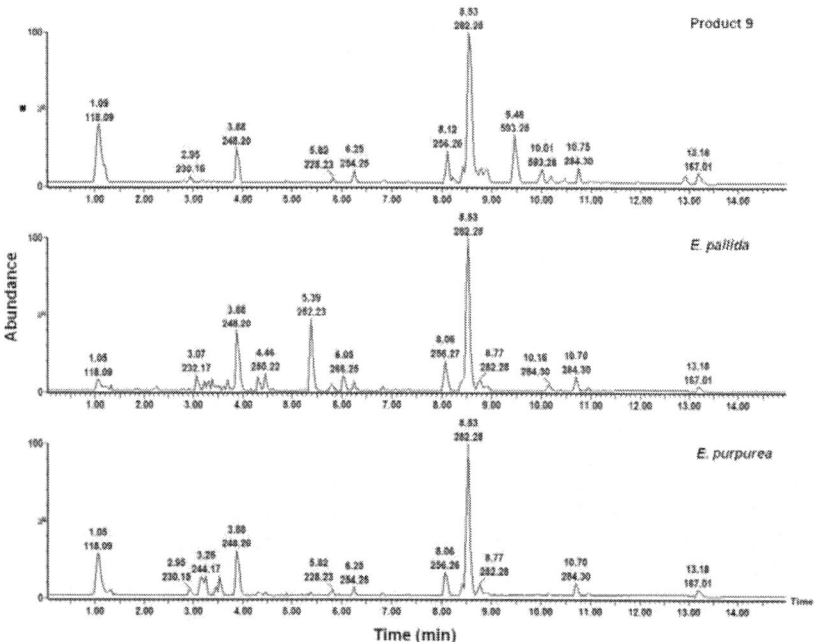

Figure 12. A comparison of the methanol root extracts of product 9 and two *Echinacea* species.

Chemical maps in the form of colour images provide a visual assessment of the composition of the raw materials and the products. Where multiple constituent species are present, it is also easy to detect. This study also highlighted labeling errors as previously reported. The outer packaging of product 13 stated that *E. pallida* was used. The same product had an inner package insert that stated that *E. purpurea* was used. This product contained 22.4% of *E. purpurea* according to HSI prediction. Product 18 purported to contain an *Echinacea* blend with all three species. The HSI results revealed 92.4% *E. purpurea*, 0.1% *E. angustifolia* and 2.0% *E. pallida*. The question then arises: can this truly be described as a blend if it contains only 0.1% and 2.0% of the species which are considered part of the blend? A blend has not been defined and it is clear that specifications for herbal products need to be set and enforced. However, in order to achieve this, good quality control methods such as HSI need to be developed and implemented.

3. EXPERIMENTAL SECTION

3.1. Sample Preparation

American Herbal Pharmacopoeia® (AHP)-certified reference material (root and leaf specimens) of *E. angustifolia*, *E. pallida* and *E. purpurea* were purchased from the American Herbal Products Association (AHPA, Silver

3. EXPERIMENTAL SECTION

Spring, MD, USA). Twenty commercial products claiming to contain *Echinacea* were purchased from local retail outlets in South Africa, as well as international suppliers. Both the botanical reference standards (roots and leaves) and products (capsule contents and tablets) were powdered to attain homogeneity using a Retsch® 400 MM ball mill (Haan, Germany) at a frequency of 30 Hz for 30 s. The powders were sieved through a 500 μm sieve (Endecotts Ltd., London, England) to ensure uniform particle size distribution and filled into plastic containers with a diameter of 8 mm. The surface of the powdered standards (in triplicate) and products were leveled with a spatula to minimise surface effects.

3.2. Short Wave Infrared (SWIR) Hyperspectral Image Acquisition

Powdered botanical standards of both the roots and leaves of three *Echinacea* species were replicated three times to include a total of 18 samples, positioned randomly on a stage, and image acquisition commenced as the samples entered the field of view. The images were acquired using a sisuChema short wave infrared (SWIR) hyperspectral pushbroom imaging system (Specim, Spectral Imaging Ltd., Oulu, Finland) with Chemadaq version 3.62.183.19 software. The system consisted of an imaging spectrograph coupled to a 2-D array mercury-cadmium-telluride (HgCdTe) detector with a light source of quartz halogen lamps mounted in reflector housing. A 50 mm high magnification lens with a spatial resolution of 0.30 μm was used to capture images at a frame rate of 100 Hz at an exposure of 3.0 ms with a spectral range of 920–2514 nm and a resolution of 6–7 nm. Images with 256 × 320 pixels and a pixel depth of 14 bits/pixel were obtained. Following image capture of the standards, twenty commercial products were positioned on the stage and imaged separately under the same conditions at room temperature. Internal dark and white reference standards were used for image calibration and to correct for variation in sample illumination.

3.3. SWIR Image Analysis

3.3.1. Principal Component Analysis (PCA)

The raw images were corrected automatically for white and dark references and converted to pseudo-absorbance (A/D converter counts to absorbance) using Evince multivariate analysis software version 2.4.0 (UmBio AB, Umeå, Sweden) [10]. Principal component analysis (PCA), which is the first step in multivariate data analysis, was applied with mean centering on the whole image (including background). PCA reduces dimensionality of large datasets by decomposing interrelated variables into a new set of coordinates (PCs) that are uncorrelated and ordered in a way that the first few PCs account for most of the variation in the data [13,14]. The greatest variance in the X data observed was between the sample and the background (the stage). The interactive score image and scatter density plots were used to remove pixels corresponding to the background, dead pixels and edge effects [10,11], revealing the variation between samples. Different mathematical pretreatment methods (standard normal variate (SNV), multiplicative scatter correction (MSC) and derivates) that minimise variability unrelated to the chemical composition of the powders were investigated for spectral pre-processing [15]. The method that provided the best separation with pertinent image differences in the score plot was chosen. The optimum number of principal components (PCs) was determined by excluding all PCs that did not significantly increase the value of the Q^2Y_{cum}. The regions at the beginning and at the end of NIR spectra are usually uninformative. Modeling these regions masks useful chemical information and hence the need to investigate these and remove them from the model so that clear chemical differences can be observed. In this study, the wavelengths ≤ 996 nm at the beginning of the spectra (920–996 nm) did not contain differentiating chemical information and it was excluded from the dataset resulting in better pixel classification. The resulting image was then assessed for chemical differences and similarities with the aim of differentiating between leaf and root samples of the different species.

3.3.2. Partial Least Squares Discriminant Analysis (PLS-DA)

PLS-DA is a pattern recognition technique that correlates variation in the X data matrix (spectral data) to an independent Y-variable (class membership), which is categorical, resulting in the discrimination or classification of samples. This is a supervised classification technique as data is assigned into classes based on prior knowledge for effective prediction of group membership in new samples. In this study, the Y-variable was generated by selecting the leaf and root images of a species in the PCA score image, and assigning these to a class (Class A: *E. angustifolia*, Class B: *E. purpurea*, Class C: *E. pallida*). Half of the samples ($n = 9$) in the PCA score image consisting of 18 samples was assigned to classes and thus comprised the calibration set while the other half ($n = 9$) formed the test set. Cross-validation of the model was performed using a random selection method. Seven rounds (iterations) of cross-validation were performed as default settings in the software. PLS factors were added to the model and the optimum number was determined by excluding factors that did not significantly increase the value of the Q^2Y_{cum} in the model. The PLS-DA score image (Y image) was then used for membership/class prediction (Min Y cut-off: 0.5; Max Y cut-off: 1.5) of the test set, mixtures and commercial samples.

3.4. Ultra Performance Liquid Chromatography-Mass Spectrometry (UPLC-MS)

Three different solvents (methanol, chloroform and hexane) were used separately for extraction of both the leaf and root materials for UPLC-MS analysis. The extracts were then dissolved in super purity methanol to obtain a concentration of 5 mg/mL before injection into the UPLC-MS. The analysis was performed on a Waters Acquity Ultra Performance Liquid Chromatographic system equipped with a PDA detector (Waters, Milford,

MA, USA). Separation was achieved on an Acquity UPLC BEH C_{18} column (150 mm × 2.1 mm, i.d., 1.7 μm particle size, Waters) maintained at 40 °C. Some preliminary tests were performed prior to setting the chromatographic conditions to obtain chromatograms with better resolution and a short analysis time. The mobile phase consisted of 0.1% formic acid (solvent A) and acetonitrile (solvent B) at a flow rate of 0.3 mL/min. The gradient elution was performed as follows: 61% A: 39% B changed to 30% A: 70% B in 2 min, to 20% A: 80% B in 4 min, to 10% A: 90% B in 4.5 min, maintained for 1 min and back to initial ratio in 1 min. The total run time was 15 min and the injection volume was 1.0 μL (full-loop injection). The positive and negative ion modes were examined and the positive ion mode provided results with more information and higher sensitivity. Thus, the mass spectrometer was operated in positive ion electrospray mode and nitrogen (N_2) was used as the desolvation gas. Data were acquired between 100 and 1000 mz^{-1}. The desolvation temperature was set to 350 °C at a flow rate of 500 Lh^{-1} and the source temperature was 100 °C. The capillary and cone voltages were set to 3500 and 35 V, respectively. Masslynx® version 4.1 software was used to process and obtain all the chromatographic data. The base peak intensity (BPI) chromatograms of the botanical reference standards and commercial products were compared to verify the species contained in the commercial products based on MS spectra and retention time. This was performed to validate the results obtained using the HSI method.

4. CONCLUSIONS

The results revealed that hyperspectral imaging in combination with multivariate data analysis could convincingly distinguish between the three *Echinacea* species. In addition, the constructed PLS-DA model with a high coefficient of determination, accurately predicted the *Echinacea* species contained in several commercially available products and identified products that did not contain crude *Echinacea* raw material. Products that contained a mixture of species were also confirmed using HSI. The method is clearly more suited for raw material identification and sensitivity was limited in the prediction of polyherbal formulations or products containing a high level of excipients. It is evident that rapid quality control methods such as developed in this study are of utmost importance to ensure high quality products. The species used should always be specified on the label and correspond to the species that is present in the product. Clearly, hyperspectral imaging is suited for the quality control of herbal medicines as it is a rapid, non-destructive method with high prediction ability, which considers a large part of the phytochemical metabolome.

ACKNOWLEDGMENTS

The authors would like to thank the Claude Leon Foundation and Tshwane University of Technology for the financial support.

AUTHOR CONTRIBUTIONS

Maxleene Sandasi: Hyperspectral imaging experimental work, UPLC-MS sample preparation and manuscript writing. Ilze Vermaak: Hyperspectral imaging experimental work and manuscript writing. Weiyang Chen: UPLC-MS experimental work and data analysis. Alvaro M. Viljoen: Conceptualization, data analysis and manuscript writing.

REFERENCES

1. Barnes, J.; Anderson, L.A.; Gibbons, S.; Phillipson, J.D. *Echinacea* species (*Echinacea angustifolia* (DC.) Hell., *Echinacea pallida* (Nutt.) Nutt., *Echinacea purpurea* (L.) Moench): A review of their chemistry, pharmacology and clinical properties. *J. Pharm. Pharmacol.* **2005**, *57*, 929–954.
2. Caruso, T.J.; Gwaltney, J.M. Treatment of the common cold with *Echinacea*: A structured review. *Clin. Infect. Dis.* **2005**, *40*, 807–810.
3. Tilburt, J.C.; Emanuel, E.J.; Miller, F.G. Does the evidence make a difference in consumer behavior? Sales of supplements before and after publication of negative research results. *J. Gen. Intern. Med.* **2008**, *23*, 1495–1498.
4. Ostrow, N. *Echinacea* shows little benefit as remedy for treating colds. Available online: http://www.bloomberg.com/news/2010-12-20/echinacea-shows-little-benefit-as-remedy-for-treating-colds-study-finds.html (accessed on 13 March 2012).
5. Barrett, B. Medicinal properties of *Echinacea*: A critical review. *Phytomedicine* **2003**, *10*, 66–86.
6. Bauer, R. Chemistry, analysis and immunological investigations of *Echinacea* phytopharmaceuticals. In *Immunomodulatory Agents from Plants*; Wagner, H., Ed.; Birkhauser Verlag: Basel, Switzerland, 1999; pp. 41–88.

7. Rininger, J.A.; Kickner, S.; Chigurupati, P.; McLean, A.; Franck, Z. Immunopharmacological activity of *Echinacea* preparations following simulated digestion on murine macrophages and human peripheral blood mononuclear cells. *J. Leukocyte Biol.* **2000**, *68*, 503–510.
8. Gowen, A.A.; O'Donnel, C.P.; Cullen, P.J.; Downey, G.; Frias, J.M. Hyperspectral imaging—An emerging process analytical tool for food quality and safety control. *Trends Food Sci. Technol.* **2007**, *18*, 590–598.
9. Gendrin, C.; Roggo, Y.; Collet, C. Pharmaceutical applications of vibrational chemical imaging and chemometrics: A review. *J. Pharm. Biomed. Anal.* **2008**, *48*, 533–553.
10. Manley, M.; du Toit, F.; Geladi, P. Tracking diffusion of conditioning water in single wheat kernels of different hardnesses by near infrared hyperspectral imaging. *Anal. Chim. Acta* **2011**, *686*, 64–75.
11. Williams, P.; Geladi, P.; Fox, G.; Manley, M. Maize kernel hardness classification by near infrared (NIR) hyperspectral imaging and multivariate data analysis. *Anal. Chim. Acta* **2009**, *653*, 121–130.
12. World Health Organization. *WHO Monographs on Selected Medicinal Plants*; World Health Organization: Geneva, Switzerland, 1999; Volume 1.
13. Geladi, P. Chemometrics in spectroscopy, Part 1. Classical chemometrics. *Spectrochim. Acta B* **2003**, *58*, 767–782.
14. Reich, G. Near-infrared spectroscopy and imaging: Basic principles and pharmaceutical applications. *Adv. Drug Deliv. Rev.* **2005**, *57*, 1109–1143.
15. Barnes, R.J.; Dhanoa, M.S.; Lister, S.J. Standard normal variate, transformation and de-trending of near-infrared diffuse reflectance spectra. *Appl. Spectrosc.* **1989**, *43*, 772–777.

CHAPTER 12

Multielemental Composition of Suet Oil Based on Quantification by Ultrawave/ICP-MS Coupled with Chemometric Analysis

Jun Jiang [1,2], *Liang Feng* [1,2], *Jie Li* [1,2], *E Sun* [1,2], *Shu-Min Ding* [1,2] *and Xiao-Bin Jia* [1,2,]*

[1] Affiliated Hospital on Integration of Chinese and Western Medicine, Nanjing University of Chinese Medicine, Xianlin Avenue 138#, Xianlin University City, Nanjing 210023, Jiangsu, China
[2] Key Laboratory of New Drug Delivery System of Chinese Meteria Medica, Jiangsu Provincial Academy of Chinese Medicine, 100# Shizi Road, Nanjing 210028, Jiangsu, China

ABSTRACT

Suet oil (SO) has been used commonly for food and medicine preparation. The determination of its elemental composition has became an important challenge for human safety and health owing to its possible contents of heavy metals or other elements. In this study, ultrawave single reaction chamber microwave digestion (Ultrawave) and inductively coupled plasma-mass spectrometry (ICP-MS) analysis was performed to determine 14 elements (Pb, As, Hg, Cd, Fe, Cu, Mn, Ti, Ni, V, Sr, Na, Ka and Ca) in SO samples. Furthermore, the multielemental content of 18 SO samples, which represented three different sources in China: Qinghai, Anhui and Jiangsu, were evaluated and compared. The optimal ultrawave digestion conditions, namely, the optimal time (35 min), temperature (210 °C) and pressure (90 bar), were screened by Box-Behnken design (BBD). Eighteen samples were successfully classified into three groups by principal component analysis (PCA) according to the contents of 14 elements. The results showed that all SO samples were rich in elements, but with significant differences corresponding to different origins. The outliers and majority of SO could be discriminated by PCA according to the multielemental content profile. The results highlighted that the element distribution was associated with the origins of SO samples. The proposed ultrawave digestion system was quite efficient and convenient, which could be mainly attributed to its high pressure and special high-throughput for the sample digestion procedure. Our established method could be useful for the quality control and standardization of elements in SO samples and products.

Keywords: Suet oil (SO); Multielements; Ultrawave digestion; ICP-MS; Chemometrics analysis

1. INTRODUCTION

Inorganic elements, important chemical components of natural foods and their products, have been found to play an important role in human safety and health. Based on emerging evidence heavy metals and metalloids, especially Pb, As, Hg and Cd, have attracted increasing attention due to their bioavailability and toxicity. These elements may be introduced into the food chain in various ways, including contamination during cultivation, processing, and storage. Other elements such as Mg, Ca, Mn, Ni, Cu, Zn, and Fe are also present and found to have both nutritional and toxic effects for human health [1,2,3].

Suet oil (SO), obtained from *Capra hircus linnaeus* or *Ovis aries linnaeus*, contains abundant inorganic elements [4]. SO can be widely applied in the food industry, health protection industry and various fields of medicine [5,6,7]. For example, SO can be used for the processing of the traditional herb Herba Epimedii. It can be assembled into smaller size nanomicelles and then promotes the absorption and enhances the clinical efficacy of the main active flavonoids [1,8,9,10]. However, inorganic elements can significantly reduce the critical micelle concentration (CMC) of the surfactant (SO) which inhibits the formation of micelles [11,12,13,14,15]. In addition, the inorganic elements in SO can have nutritional functions or toxic effects for human health. Therefore, the quality control of inorganic elements in SO can ensure its safety and function, and also provide important data for the use of SO in the food and medicine industries.

Inductively coupled plasma mass spectrometry (ICP-MS) is a powerful technique for elemental analysis in various fields [16,17]. Before ICP-MS instrument analysis, sample digestion efficiency is a critical pretreatment step affecting analytical results, especially for oleaginous matrices [18], such as SO. Microwave-assisted digestion including closed vessel systems has been widely used as an original way to pretreat samples for ICP-MS analysis [19,20]. The ultrawave single reaction chamber microwave digestion system overcame the limitations of traditional microwave sample preparation. At the heart of this system, there is a large sample chamber which is pre-pressurized with an inert gas and heated with microwave energy. The chamber serves both as a microwave cavity and a reaction vessel. Different samples can be digested simultaneously because no vessel assembly/disassembly is required while vessel cleaning is eliminated with the use of disposable glass vials. Direct temperature and pressure control of every sample ensures complete control of the digestion process. Blanks are significantly lower than with closed vessel digestion, since less solution transfer occurs, quartz vials can be used and digestion acid volumes are lower.

Although it has been used widely in the food and medicine industries, to the best of our knowledge, the standardization of SO still remains a challenge due to contamination or purposeful adulteration with other oils. It is difficult to identify the origins and species of SO based on their appearance and morphology. In this study, eighteen batches of SO were collected from three origins. Ultrawave digestion parameters (digestion time, digestion temperature, digestion pressure) were optimized by Box-Behnken Design (BBD) statistical screening and applied to the pretreatment of SO for the subsequent determination of Pb, As, Hg, Cd, Fe, Cu, Mn, Ti, Ni, V, Sr, Na, Ka and Ca by ICP-MS. Finally, principal component analysis (PCA) was performed to evaluate and classify the eighteen batches of SO according to the detected contents of the various elements.

2. RESULTS AND DISCUSSION

2.1. Optimization for SO Ultrawave Digestion by Box-Behnken Design

Ultrawave digestion can be used to save consumables costs and sample pretreatment time owing to its higher performance and throughput. In addition, ultrawave digestion could be operated up to 199 bar pressure and 240 °C temperature, which can easily digest oleaginous matrices. Box-Behnken design (BBD), a collection of mathematical and statistical techniques, was first established by Box and Wilson [21]. It has been widely applied for improving or optimizing processing conditions in food and pharmaceutical studies [22].

In this study, one batch of SO sample obtained from Jiangsu was used to optimize the digestion conditions by BBD (Figure 1), and its blank and spiked samples were pretreated in parallel three times. Then the recovery was calculated as follows:

[(Measured value − Original value)/Spiked value] × 100%.

Finally, the optimal conditions for digestion were screened out by BBD. The ultrawave digestion procedure is shown in Table 1. Operating conditions and parameters for the ICP-MS and ultrawave single reaction chamber microwave digestion systems are shown in Table S1. As shown in Table S2, three digestion variables including temperature (A), time (B) and pressure (C) were examined by BBD design. In this optimization, the recoveries of 14 metal elements were chosen as the response. Comprehensive test results of response surface plots (3D) showed that the recoveries of 14 elements were all higher than 80% under the experimental conditions of temperature at 210 °C, pressure at 90 bar and digestion time for 35 min.

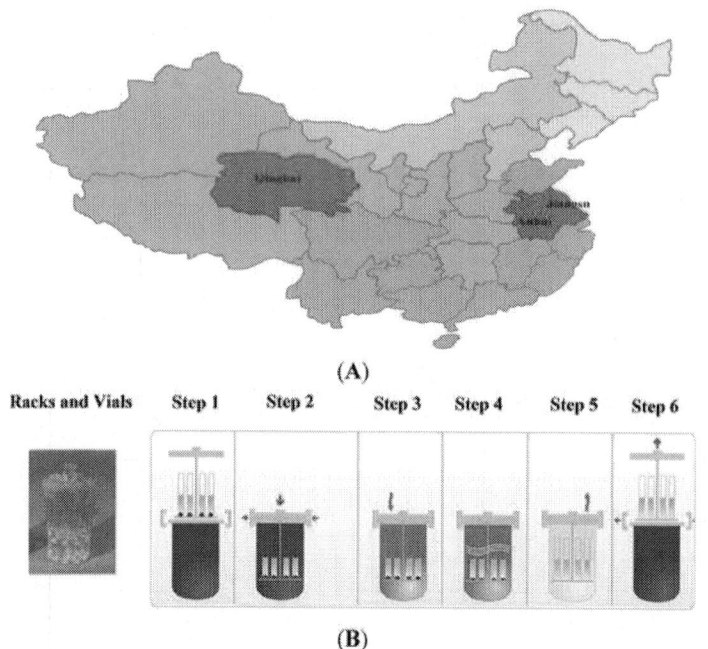

Figure 1. Origins of the samples and the ultrawave digestion procedure. (**A**) The distribution of the 18 batches of SO in China; (**B**) The ultrawave digestion procedure.

Table 1. Ultrawave digestion procedure.

Steps	Status	Temperature/°C	Time/min
1	1600 W heating	25–120	5
2	Insulation	120	5
3	1600 W heating	120–210	5
4	Insulation	210	35

Precharged nitrogen pressure of 90 bar.

The experimental data was analyzed by ANOVA and the results are given in Table 2 and Table S2. Values of "prob > F" less than 0.001 indicated that the model terms are significant. R-Squared higher than 0.99 for less than 1.0% of the total variations, indicated that the optimal results were accurate and reliable. Therefore, we could find out the optimal condition from (3D) response surface plots was a temperature at 210 °C while the pressure was kept at 90 bar for 35 min (Figure 2). The correlation coefficient of 14 elements were all at 0.99 under ICP-MS conditions, while the standard curves were forced through the origin. Blank sample was repeatedly testing 10 times, and three times the standard deviation was the limit of detection (LOD), and 10 times the standard deviation was the limit of quantitation (LOQ). The LOD were 0.002–0.6 μg/kg, LOQ were 0.335–9.1 μg/kg, the recovery rates of 14 standards varied from 86.5%–99.2% (Table 3), revealing the accuracy of this method.

2. RESULTS AND DISCUSSION

Table 2. ANOVA for response surface model.

Response	Final Equation	Std. Dev.	Mean	C.V. %	PRESS	R-Squared
Pb	Pb = 85.72 + 3.82A + 4.94B + 10.24C − 0.97AB + 2.28AC + 1.05BC − 5.03A^2 − 4.68B^2 − 4.79C^2	6.32	75.81	8.34	2913.15	0.9021
As	As = 83.56 + 4.29A + 2.96B + 11.96C + 1.69AC − 1.36AB − 0.44BC − 3.30A^2 − 1.20B^2 − 5.46C^2	5.57	76.76	7.26	2152.43	0.9035
Hg	Hg = 84.15 + 2.78A + 2.02B + 13.83C + 0.15AB + 1.48AC − 1.62BC − 3.74A^2 − 2.89B^2 − 5.56C^2	4.95	75.81	6.52	1632.57	0.9341
Cd	Cd = 82.92 + 3.13A + 2.37B + 13.21C + 0.30AB + 1.88AC − 1.27BC − 3.40A^2 − 3.58B^2 − 5.07C^2	5.88	74.69	7.87	2239.73	0.9038
Fe	Fe = 81.10 + 1.80A + 1.80B + 11.51C + 1.20AB + 2.23AC + 0.53BC − 1.48A^2 − 3.25B^2 − 4.19C^2	4.58	75.01	6.10	1054.59	0.9175
Cu	Cu = 83.23 + 3.44A + 4.06B + 10.68C − 1.36BC + 2.51AC − 1.36BC − 3.46A^2 − 2.27B^2 − 4.25C^2	4.77	76.42	6.25	955.60	0.9151
Mn	Mn = 85.88 + 5.23A + 4.67B + 10.41C − 0.30AB + 1.60AC − 2.02BC − 4.09A^2 − 4.20B^2 − 2.82C^2	5.40	78.30	6.90	1939.88	0.9031
Ti	Ti = 86.85 + 5.07A + 3.30B + 8.45C + 1.96AB + 1.96AC + 3.11BC − 4.23A^2 − 3.47B^2 − 2.97C^2	3.32	79.56	4.17	253.82	0.9498
Ni	Ni = 89.17 + 7.35A + 3.41B + 9.18C + 0.89AB + 3.39AC − 1.64BC − 6.99A^2 − 3.68B^2 − 2.36C^2	3.92	80.27	4.89	726.98	0.9517
V	V = 87.65 + 7.26A + 2.41B + 9.21C + 0.85AB + 6.05AC − 3.25BC − 5.74A^2 − 2.96B^2 − 4.18C^2	3.23	78.85	4.10	789.74	0.9671
Cr	Cr = 86.95 + 5.05A + 1.37B + 8.23C + 2.34AB + 2.14AC − 0.39BC − 6.88A^2 − 2.58B^2 − 3.33C^2	6.02	78.22	7.70	2293.24	0.9060
Na	Na = 73.30 + 2.44A + 4.97B + 10.24C − 0.075AB + 3.30AC + 2.08BC − 10.32A^2 + 0.30B^2 − 4.91C^2	5.70	63.11	9.04	2112.79	0.9205
K	K = 71.15 + 0.54A + 2.90B + 9.96C − 2.14AB + 3.66AC + 2.99BC − 9.80A^2 + 0.29B^2 − 5.65C^2	6.64	60.80	10.92	3005.07	0.9062
Ca	Ca = 78.15 + 1.02A + 0.82B + 8.36C + 1.79AB − 2.49AC + 2.54BC − 7.43A^2 − 1.53B^2 − 4.41C^2	4.92	69.03	7.12	1449.21	0.9176

A, Time; B, Temperature; C, Pressure.

Table 3. The linear regression equations, the correlation coefficient (r), method detection limits(LOD, LOQ), precision, repeatability, and recovery of 14 elements under ICP-MS conditions.

Elements	Linear equation	Linearity range µg/L	r	LOD µg/kg	LOQ µg/kg	Precision (RSD, n = 6) %		Repeatability (RSD, n = 6) %	Recovery (%)
						Intraday	Interday		
Pb	Y = 66211X	0–10	0.9998	0.005	0.353	1.23	1.56	2.36	97.8
As	Y = 14879X	0–5000	1.0000	0.02	2.56	2.31	2.43	2.07	98.6
Hg	Y = 6572.3X	0–10	1.0000	0.004	0.708	1.43	1.63	3.24	96.4
Cd	Y = 3955.5X	0–10	1.0000	0.002	0.335	2.03	2.34	2.78	98.3
Fe	Y = 215.19X	0–5000	0.9981	0.09	3.64	1.78	2.44	2.78	97.4
Cu	Y = 14242 X	0–10	0.9996	0.008	3.091	2.58	1.96	3.35	88.7
Mn	Y = 7966.1X	0–50	1.0000	0.02	5.18	2.32	2.55	2.23	93.8
Ti	Y = 168.39X	0–500	0.9993	0.2	5.8	1.73	2.05	2.19	86.5
Ni	Y = 10693X	0–50	0.9982	0.009	5.668	1.99	2.57	2.47	90.4
V	Y = 9399.3X	0–10	1.0000	0.003	2.608	2.27	2.79	2.94	96.5
Cr	Y = 1441.2X	0–50	1.0000	0.002	6.088	1.66	2.16	2.38	99.2
Na	Y = 45249X	0–5000	0.9988	0.2	9.1	1.07	2.76	3.76	89.7
K	Y = 15646X	0–5000	0.9979	0.6	8.1	2.06	2.37	2.89	90.2
Ca	Y = 51.87X	0–500	0.9975	0.3	5.1	1.65	2.39	3.25	87.9

2.2. Application to Multielemental Analysis in SO

In this study, fourteen elements in SO samples were successfully determined to compare each one. Under our experimental conditions, the digested solutions containing of the fourteen elements, including Pb, As, Hg, Cd, Fe, Cu, Mn, Ni, Ti, V, Cr, Na, K and Ca, were directly injected and analyzed after constant volume.

As shown in Table 4, an obvious difference in the concentrations of all 14 elements was observed in different SO samples. The contents of elements detected in various samples were in the range from 0.0015 to 1.4 mg/kg for Pb, from 0.013 to 784 mg/kg for As, from 0.0015 to 0.025 mg/kg for Hg, from 0.239 to 1.61 mg/kg for Cu, from 196 to 817 mg/kg for Na, from 94 to 584 mg/kg for K, from 18 to 80 mg/kg for Ca, from 3.81 to 21 mg/kg for Ti, from 0.079 to 0.274 mg/kg for V, from 4.00 to 9.81 mg/kg for Cr, from 0.61 to 2.03 mg/kg for Mn, and from 71.6 to 224 mg/kg for Fe.

Among the analyzed elements, Na was the most abundant element, followed by K, Fe, Ca, Ti, Cr, Ni, Mn, Cu, V, As, Pb, Hg, and Cd. In particular, the content of elements such as Pb, As, Hg, Cd, V were very low. However, a high amount of As was found in batches 5 and 6 (784 mg/kg and 561 mg/kg), which might have been contaminated during preparation or storage. Concentrations of Na, K, Ca, and Fe were more higher than other elements in the SO, especially in batches 1, 6, 7, and 8. Unexpectedly, Cd and Ni were not detected in batches 1–4, 14, and 15, 18, respectively.

2. RESULTS AND DISCUSSION

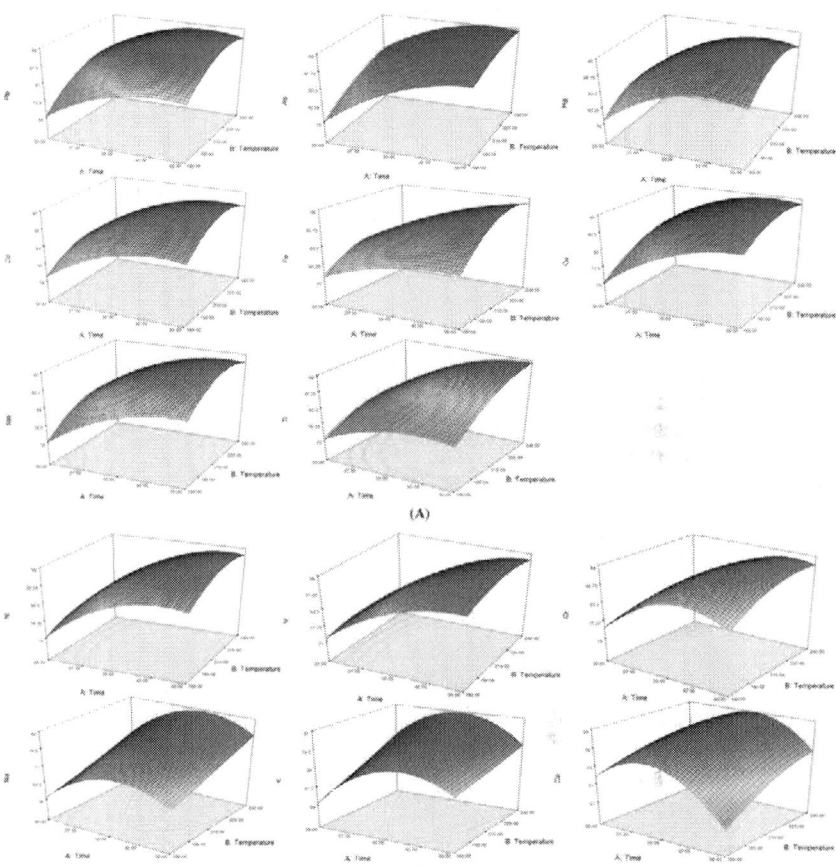

Figure 2. Optimization of ultrawave digestion parameters by BBD. (**A**) Response surface plots (3 D) of Pb, As, Hg, Cd, Fe, Cu, Mn, Ti; (**B**) Response surface plots (3 D) of Ni, V, Cr, Na, K, Ca.

2.3. PCA of the SO Samples

To evaluate the variation of SO, PCA was performed on the basis of the contents of 14 tested elements from SO. The first three principal components (PC1, PC2, and PC3) with >84.6% of the whole variance were extracted for analysis. Among them, PC1 accounted for 51.49% of total variance, whereas PC2 and PC3 for 22.57% and 10.58%, respectively. The remaining principal components were discarded due to their minor effects on the model. The correlation matrix of PCA analysis for 14 elements is shown in Table S3. The total variance for PCA and the components loading matrix is shown in Table 5 and Table 6. According to their loadings, PC1 had good correlation with the 14 elements. The above results suggested that most of the elements might contribute to the classification of the samples.

Table 4. Contents of 14 types elements in 18 batchs of suet oil (n = 3).

Origins	Batches	Pb	As	Cd	Hg	Cu	Na	K	Ca	Ti	V	Cr	Mn	Fe	Ni
Anhui	1	0.0015±0.0008	0.015±0.003	nd	0.004±0.001	0.82±0.01	817±4	584±5	80±4	21±1	0.274±0.007	9.81±0.03	2.03±0.05	224±5	2.45±0.06
	2	0.18±0.01	0.095±0.006	nd	0.0040±0.0003	1.11±0.09	386±5	240±6	29±1	7.8±0.1	0.14±0.01	8.94±0.04	1.03±0.02	148±6	1.02±0.05
	3	0.20±0.01	0.035±0.007	nd	0.0030±0.0001	1.45±0.02	411±6	314±8	34±4	8.07±0.04	0.15±0.01	8.0±0.1	1.16±0.01	145±3	1.02±0.01
	4	0.11±0.01	0.026±0.004	nd	0.0020±0.0001	0.43±0.01	323±3	206±6	24±2	1.60±0.06	0.113±0.006	7.90±0.05	0.816±0.008	120±5	0.94±0.01
	5	0.70±0.03	784±8	0.0132±0.001	0.0002±0.0001	0.276±0.004	360±4	214±5	26±3	8.79±0.07	0.150±0.004	7.9±0.1	1.02±0.01	179±7	0.593±0.008
	6	0.13±0.01	561±11	0.0010±0.0003	0.0015±0.0001	0.239±0.003	632±4	444±9	39±4	19.4±0.1	0.172±0.003	4.07±0.05	1.88±0.02	181±4	3.79±0.07
Qingdao	7	1.4±0.1	0.34±0.02	0.030±0.006	0.025±0.001	1.34±0.02	580±7	392±7	50±2	11.61±0.01	0.194±0.002	7.72±0.03	1.36±0.02	157±4	1.42±0.02
	8	0.8±0.1	0.26±0.02	0.016±0.002	0.015±0.001	0.90±0.02	703±8	488±7	59±3	18.57±0.08	0.162±0.004	4.00±0.06	1.68±0.02	188±2	2.38±0.02
	9	1.3±0.1	0.31±0.02	0.020±0.003	0.020±0.001	1.21±0.01	291±5	147±5	26±1	5.76±0.08	0.084±0.005	4.356±0.002	0.91±0.02	93.1±0.2	0.59±0.03
	10	1.4±0.1	0.30±0.02	0.027±0.003	0.018±0.001	1.49±0.02	320±5	159±5	29±3	4.82±0.07	0.224±0.009	6.570±0.002	0.91±0.01	151±3	2.52±0.02
	11	1.2±0.1	0.275±0.008	0.021±0.001	0.0165±0.0004	1.606±0.006	426±5	253±6	41±3	3.2±0.1	0.143±0.001	7.074±0.006	1.04±0.01	126±2	0.99±0.01
Jiangsu	12	0.9±0.1	0.104±0.009	0.013±0.001	0.013±0.001	0.844±0.006	296±8	170±6	27±2	6.17±0.04	0.1035±0.0008	7.12±0.04	0.79±0.01	90.9±0.8	2.74±0.03
	13	0.8±0.1	0.04±0.01	0.003±0.001	0.0095±0.0002	0.389±0.009	308±6	170±6	22±1	4.95±0.04	0.099±0.008	7.138±0.013	0.72±0.03	89±2	0.866±0.04
	14	1.2±0.1	0.226±0.006	nd	0.023±0.002	0.90±0.01	246±6	127±7	23±1	4.16±0.03	0.095±0.002	6.08±0.04	0.73±0.05	83±1	0.52±0.02
	15	0.8±0.1	0.21±0.02	0.0055±0.0003	0.0065±0.0002	0.586±0.007	196±5	94±2	18±2	4.55±0.04	0.079±0.003	5.15±0.03	0.62±0.06	79±2	nd
	16	0.7±0.1	0.212±0.008	0.0030±0.0008	0.0090±0.0008	0.33±0.01	264±5	122±3	22±2	3.81±0.02	0.092±0.002	5.96±0.04	0.62±0.01	83.8±0.5	0.134±0.002
	17	0.70±0.07	0.12±0.01	0.0030±0.0008	0.0085±0.0003	0.756±0.007	301±6	141±2	28±2	4.10±0.04	0.094±0.004	3.98±0.06	0.755±0.006	85.8±0.4	1.22±0.01
	18	0.91±0.08	0.041±0.006	0.006±0.001	0.0095±0.0004	0.620±0.004	294±4	141±3	22±2	4.0±0.2	0.086±0.002	3.92±0.02	0.61±0.01	71.6±0.8	nd

nd, not detected (concentration below the LOD).

2. RESULTS AND DISCUSSION

Table 5. The total variance explained for PCA of 14 elements in 18 batches of SO.

Component	Initial Eigenvalues			Extraction Sums of Squared Loadings		
	Total	% of Variance	Cumulative %	Total	% of Variance	Cumulative %
1	7.209	51.491	51.491	7.209	51.491	51.491
2	3.160	22.572	74.062	3.160	22.572	74.062
3	1.482	10.586	84.649	1.482	10.586	84.649
4	0.951	6.794	91.443			
5	0.479	3.423	94.866			
6	0.301	2.153	97.019			
7	0.179	1.280	98.299			
8	0.134	0.960	99.259			
9	0.059	0.418	99.677			
10	0.019	0.133	99.810			
11	0.014	0.102	99.913			
12	0.008	0.060	99.973			
13	0.003	0.018	99.991			
14	0.001	0.009	100.000			

Extraction Method: Principal Component Analysis.

Table 6. The component matrix of PCA analysis for 14 elements in 18 batches of SO.

Elements	Component		
	1	2	3
Pb	−0.440	0.825	0.251
As	0.294	−0.400	0.563
Cd	0.032	0.869	0.205
Hg	−0.232	0.902	0.108
Cu	0.018	0.776	−0.469
Na	0.966	0.112	0.001
K	0.970	0.057	−0.042
Ca	0.955	0.152	−0.005
Ti	0.965	−0.012	0.166
V	0.851	0.253	−0.220
Cr	0.226	−0.159	−0.825
Mn	0.982	0.055	0.089
Fe	0.939	0.023	−0.039
Ni	0.732	0.132	0.244

Extraction Method: Principal Component Analysis.

In order to further distinguish the diversity of SO samples from different origins, the scatter plot of the study was plotted. We observed that eighteen sample dots were successfully classified into groups I, group II, and group III corresponding to Qinghai, Anhui and Jiangsu (Figure 3). Interestingly, dots in groups II and III were relatively nearer to each other, indicating a closer relationship among six batches from Anhui and seven batches from Jiangsu.

However, dots in group I were relatively scattered, suggesting the diversification of the five Qinghai batches. This could be explained by several reasons: firstly, the land area of Qinghai Province is 722,300 square kilometers, larger than Anhui (139,600) and Jiangsu (106,700 square kilometers), which creates advantageous conditions for the diversity of the samples. Secondly, the climate and environment in Qinghai, Anhui and Jiangsu have great differences, which affects the differences in elemental metabolism in *Ovis aries Linnaeus* or *Capra hircus Linnaeus*. Thirdly, Anhui Province and Jiangsu Province are neighboring to each other in geographic location, which is the main reason for the closer results of the samples from the two origins.

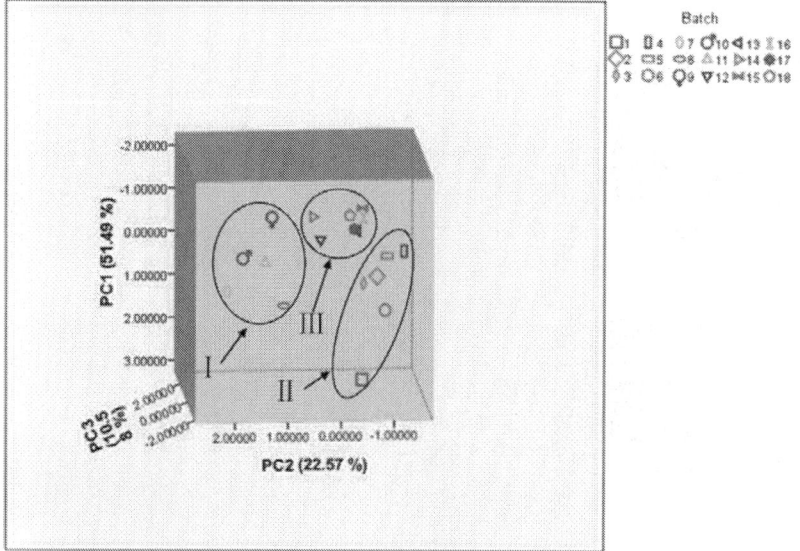

Figure 3. The 3 D scatter plots obtained from PCA of 18 batches SO samples.

3. EXPERIMENTAL

3.1. General Information

The analytical method used in this study is based on an ultrawave single reaction chamber microwave digestion system (Ultrawave) and inductively coupled plasma-mass spectrometry (ICP-MS), and consists of the following steps: (1) optimization of ultrawave digestion by Box-Behnken design (BBD); (2) measurement of 14 elements in 18 batches of SO by ICP-MS; (3) differentiation of 18 batches of SO by principal component analysis.

3.2. Instrumentation

An inductively coupled plasma quadrupole mass spectrometer (Agilent 7700, Agilent, Santa Clara, CA, USA) coupled with an automatic sampler (ASX-500,

Agilent) was used in all experiments. The ultrawave single reaction chamber microwave digestion system (Milestone, Milan, Italy) including a 1 L stainless steel reaction chamber and vials and a rack was applied for all SO sample digestions (Table 1 and TableS1).

3.3. Reagents and Analytical Solutions

Working standard solutions of 14 elements (Pb, As, Hg, Cd, Fe, Cu, Mn, Ti, Ni, V, Sr, Na, Ka and Ca) were prepared by diluting proper amounts of their inorganic standard solutions (1,000 mg L^{-1} each, Kanto Chemical Industries, Ltd., Tokyo, Japan). Nitric acid and hydrogen peroxide were of ultrapure grade (Kanto Chemical). Pure water (resistivity 18 MΩ cm) was prepared by a Milli-Q water purification system (Millipore Corp., Bedford, MA, USA) and used throughout the experiments.

3.4. Samples

In total, 18 batches SO samples were purchased in the Qinghai, Jiangsu and Anhui provinces of China from January 2013 to July 2013, respectively (Figure 1). SO samples of Qinghai were derived from *Capra hircus Linnaeus*, and the samples of Jiangsu and Anhui were obtained from *Ovis aries Linnaeus*. In order to prevent oxidation, all SO samples were stored in a dark environment and kept below 4 °C in a refrigerator.

3.5. Sample Preparation Procedure

The SO samples (100 mg) were weighted accurately and placed into vials respectively, and then HNO_3 (69%, 1.5 mL) and H_2O_2 (30%, 0.5 mL) were added. Teflon caps on the glass vials were kept loose to ensure pressure equalization. Sample rack configurations including a 15 position rack for microsamples was used. The temperature and pressure of the ultrawave digestion were maintained at optimal conditions. Sequentially, the digestion procedure was divided into the following five steps: (1) sample rack was lowered automatically into microwave chamber; (2) chamber clamp is secured by the operator. Interlocks prevent operation without this clamp in place; (3) chamber is pre-pressurized with an inert gas to prevent sample boiling. Cross contamination is eliminated; (4) m icrowave energy is applied. All samples are under same temperature and pressure conditions; (5) very fast cooling step due to water cooling of chamber. Chamber is vented and acid vapors extracted; (6) clamp is released and the sample rack automatically rises from chamber. Figure 1 shows how a digestion was performed in the Ultrawave apparatus. After cooling the vials, the solution in the vessel was transferred and made up to volume in a 25 mL volumetric flask. A blank was also prepared in the same manner.

3.6. PCA for SO Samples

PCA was performed to analyze these samples using the SPSS 16.0 software package (SPSS, Chicago, IL, USA). In this study, the contents of the 14 elements which were analyzed from the 18 batches SO samples composed a data matrix with 18 rows and 14 columns, which was used for PCA analysis after normalization. The first three principal components (PCs) were extracted, and the scatter plots were obtained by plotting the scores of PC1 *versus* PC2 and PC3.

4. CONCLUSIONS

In this paper, 18 batches of SO samples were analyzed by Ultrawave digestion/ICP-MS with high efficiency, high throughput, lower labor and high accuracy. In addition, the combination of the concentration of multielements with PCA was also a reliable means for the detection and differentiation of trace constituents, which could be used to provide supportive information on inorganic elements contents for evaluating the quality of SO and its products. In summary, the present approach is effective, reliable and can be applied for evaluating the safety and nutritional function of SO and other related products with undefined quality control standards.

ACKNOWLEDGMENTS

This work was supported by the Natural Science Foundation of China (Nos. 81274088), the Science and Technology Support Program of Zhenjiang City—Social Development Project (SH2012010).

AUTHOR CONTRIBUTIONS

JJ and XBJ designed research; JJ, JL, SMD, ES and LF performed research and analyzed the data; JJ and LF wrote the paper. All authors read and approved the final manuscript.

REFERENCES

1. Raman, P.; Patino, L.C.; Nair, M.G. Evaluation of metal and microbial contamination in botanical supplements. *J. Agric. Food Chem.* **2004**, *52*, 7822–7827.
2. Ong, E.S.; Yong, Y.L.; Woo, S.O. Determination of lead in botanicals/Chinese prepared medicines by using microwave digestion with flow injection-inductively coupled plasma-mass spectrometry. *J. AOAC Int.* **2000**, *83*, 382–389.
3. López, A.M.; Prieto, M.F.; Miranda, M.; Castillo, C.; Hernández, J.; Luis, B.J. Interactions between toxic (As, Cd, Hg and Pb) and nutritional essential (Ca, Co, Cr, Cu, Fe, Mn, Mo, Ni, Se, Zn) elements in the tissues of cattle from NW Spain. *Biometals* **2004**, *17*, 389–397.

4. Perelló, G.; Martícid, R.; Llobet, J.M.; Domingo, J.L. Effects of various cooking processes on the concentrations of arsenic, cadmium, mercury, and lead in foods. *J. Agric. Food Chem.* **2008**, *56*, 11262–11269. [
5. Lu, Y.; Wang, J.; Deng, Z.; Wu, H.; Deng, Q.; Tan, H.; Cao, L. Isolation and characterization of fatty acid methyl ester (FAME)-producing Streptomyces sp. S161 from sheep (Ovis aries) faeces. *Lett. Appl. Microbiol.* **2013**, *57*, 200–205.
6. Mattacks, C.A.; Sadler, D.; Pond, C.M. Site-specific differences in the action of NRTI drugs on adipose tissue incubated *in vitro* with lymphoid cells, and their interaction with dietary lipids. *Comp. Biochem. Physiol. C Toxicol. Pharmacol.* **2003**, *135*, 11–29.
7. Pond, C.M.; Mattacks, C.A. The source of fatty acids incorporated into proliferating lymphoid cells in immune-stimulated lymph nodes. *Br. J. Nutr.* **2003**, *89*, 375–383.
8. Cui, L.; Sun, E.; Zhang, Z.H.; Tan, X.B.; Wei, Y.J.; Jin, X.; Jia, X.B. Enhancement of epimedium fried with suet oil based on *in vivo* formation of self-assembled flavonoid compound nanomicelles. *Molecules* **2012**, *17*, 12984–12996.
9. Editorial Committee of Pharmacopoeia of Ministry of Health PR China. *The Pharmacopeoia of People's Republic of China*; China Chemical Industry Press: Beijing, China, 2010; pp. 306–308.
10. Vanhasselt, P.M.; Janssens, G.E.; Slot, T.K.; Vander, H.M.; Minderhoud, T.C.; Talelli, M.; Akkermans, L.M.; Rijcken, C.J.; van Nostrum, C.F. The influence of bile acids on the oral bioavailability of vitamin K encapsulated in polymeric micelles. *J. Control. Release* **2008**, *133*, 161–168.
11. Jiang, L.; Wang, K.; Deng, M.; Wang, Y.; Huang, J. Bile salt-induced vesicle-to-micelle transition in catanionic surfactant systems: Steric and electrostatic interactions. *Langmuir* **2008**, *24*, 4600–4606.
12. Shen, X.; Huang, W.; Yao, C.; Ying, S. Influence of metal ion on sorption of p-nitrophenol onto sediment in the presence of cetylpyridinium chloride. *Chemosphere* **2007**, *67*, 1927–1932.
13. Das, P.; Mallick, A.; Sarkar, D.; Chattopadhyay, N. Application of anionic micelle for dramatic enhancement in the quenching-based metal ion fluorosensing. *J. Colloid Interface Sci.* **2008**, *320*, 9–14.
14. Gholivand, M.B.; Babakhanian, A.; Rafiee, E. Determination of Sn(II) and Sn(IV) after mixed micelle-mediated cloud point extraction using alpha-polyoxometalate as a complexing agent by flame atomic absorption spectrometry. *Talanta* **2008**, *76*, 503–508.
15. Mu, J.H.; Li, G.Z. The formation of wormlike micelles in anionic surfactant aqueous solutions in the presence of bivalent counterion. *Chem. Phys. Lett.* **2001**, *345*, 100–104.

16. Vonderheide, A.P.; Sadi, B.B.M.; Sutton, K.L.; Shann, J.R.; Caruso, J.A. *Inductively Coupled Plasma Spectrometry and Its Application*; Blackwell Publishing Ltd.: Oxford, UK, 2007; pp. 338–361. [Google Scholar]
17. Taylor, H.E.; Huff, R.A.; Montaser, A. *Inductively Coupled Plasma Spectrometry*; Wiley-VCH Inc.: Danvers, MA, USA, 1998; pp. 721–765.
18. Sucharová, J.; Suchara, I. Determination of 36 elements in plant reference materials with different Si contents by inductively coupled plasma mass spectrometry: comparison of microwave digestions assisted by three types of digestion mixtures. *Anal. Chim. Acta* **2006**, *576*, 163–176.
19. Abusamra, A.; Morris, J.S.; Koirtyohann, S.R. Wet ashing of some biological samples in a microwave oven. *Anal. Chem.* **1975**, *47*, 1475–1477.
20. Suoranta, T.; Niemelä, M.; Perämäki, P. Comparison of digestion methods for the determination of ruthenium in catalyst materials. *Talanta* **2014**, *119*, 425–429.
21. Box, G.E.P.; Wlson, K.B. On the experimental attainment of optimum conditions. *J. Roy. Statist. Soc. Ser. B Metho.* **1951**, *13*, 1–45.
22. Luo, C.; Chen, Y.S. Optimization of extraction technology of Se-enriched *Hericium erinaceum* polysaccharides by Box–Behnken statistical design and its inhibition against metal elements loss in skull. *Carbohydr. Polym.* **2010**, *82*, 845–860.

CHAPTER 13

Chemometric Feature Selection and Classification of Ganoderma lucidum Spores and Fruiting Body Using ATR-FTIR Spectroscopy

Ying Zhu[1*], Augustine Tuck Lee Tan[2]

[1]Mathematics and Mathematics Education, National Institute of Education, Nanyang Technological University, Singapore
[2]Natural Sciences and Science Education, National Institute of Education, Nanyang Technological University, Singapore

ABSTRACT

Ganoderma lucidum (G. lucidum) spores as a valuable Chinese herbal medicine have vast marketable prospect for its bioactivities and medicinal efficacy. This study aims at the development of an effective and simple analytical method to distinguish G. lucidum spores from its fruiting body, which is of essential importance for the quality control and fast discrimination of raw materials of Chinese herbal medicine. Attenuated total reflection Fourier transform infrared (ATR-FTIR) spectroscopy combined with the appropriate chemometric methods including penalized discriminant analysis, principal component discriminant analysis and partial least squares discriminant analysis has been proven to be a rapid and powerful tool for discrimination of G. lucidum spores and its fruiting body with classification accuracy of 99%. The model leads to a well-performed selection of informative spectral absorption bands which improve the classification accuracy, reduce the model complexity and enhance the quantitative interpretations of the chemical constituents of G. lucidum spores regarding its anticancer effects.

Keywords: Feature Selection, Attenuated Total Reflection Fourier Transform Infrared Spectroscopy, Penalized Linear Discriminant Analysis, Principal Component Discriminant Analysis, Partial Least Squares Discriminant Analysis

1. INTRODUCTION

Ganoderma lucidum (G. lucidum), a fungus famous as traditional Chinese herbal medicine, has been widely used for preventing and treating a series of diseases. G. lucidum spores are the fungus's reproductive cells ejected from the cap of G. lucidum after its fruiting body becomes mature. Though the fruiting body of G. lucidum has been widely utilized as a Chinese medicine for several thousand years, the spores of G. lucidum have been realized and utilized only since the 20th century. Recent studies demonstrated that the spores of G. lucidum not only inherit all active ingredients of G. lucidum, but also have stronger bioactivities, about 75 times more than G. lucidum's fruiting body regarding its effect, such as enhancing immunity, antitumor, preventing diabetes, protecting liver and so on [1] [2] . As G. lucidum spores are valuable Chinese herbal medicine, its potent bioactivity and wide acceptability make it an important marketable product. However, the market product from fruiting body of G. lucidum sometimes pretends to be product from G. lucidum spores for economic benefit. Therefore, it is of great importance to develop an accurate, efficient and simple method to identify G. lucidum spores from its fruiting body since it is very difficult to discriminate them from their market products.

G. lucidum spores have a complicated system of compounds. The commonly investigated methods for the analysis of herbal medicines, like high performance liquid chromatography (HPLC), thin layer chromatography (TLC) and colorimetry, are found to be expensive, time-consuming, labour-intensive, and requiring large quantity of organic solvents. Also, the results are inadequate for classification purpose because of the limited amount of active chemical components that can be detected in what is a very complex system in G. lucidum spores [3] [4].

Attenuated total reflection Fourier transform infrared (ATR-FTIR) spectroscopy is very efficient for revealing chemical compositions and structures of herbal medicine in terms of easy and direct usage of technique, non- destructiveness, small quantity of sample needed and short data acquisition time. Since chemical processing of the herbal material is not needed at all, the chemical composition of the material remains in its original form [5]. However, studies on herbal medicines using the ATR-FTIR technique are still in its early stages [6] and there were few studies on ATR-FTIR spectroscopy of G. lucidum spores.

ATR-FTIR spectra of herbal medicines consist of many overlapping absorption bands representing the different modes of vibration of a large number of molecular constituents in the compounds. These vibrational bands are sensitive to the physical and chemical states of the compounds, and they can be detected at low levels [4]. However, the differences in the ATR-FTIR spectra within the same herbal species may be subtle and it is difficult to distinguish between spores and fruiting body of G. lucidum through simple visual inspection. Thus, approriate multivariate chemometric methods have been applied in this study to analyze the ATR-FTIR spectra for feature selection and classification. Although the spectroscopic data are highly correlated, often, only a small subset of spectral features is found essential. Due to the complex nature of the combined spectroscopic data, it is of great interest to determine if a small subset of spectral reflectance measurements contains as much informative feature for classification purpose as the whole spectrum does. There is little scientific research paper available regarding the spectral feature and chemical composition of G. lucidum spores.

In this paper a penalized discriminant analysis (PDA) model was developed to identify informative spectral features for distinguishing between spores and fruiting body of G. lucidum. The multivariate methods using principal component discriminant analysis (PCDA) and partial least squares discriminant analysis (PLSDA) were also explored based on the whole spectrum. The model performances based on the selected wavelength bands and the whole spectrum were compared in terms of classification accuracy and interpretation of spectral features. The established discrimination models explored in this paper would be an accurate, simple and robust tool for the quality control of G. lucidum spores. In particular, the discriminant vectors can be helpful for the interpretation of spectral features needed for discrimination and for providing a quantitative explanation of the major chemical constituents of G. lucidum spores regarding its anti-cancer effects.

2. MATERIALS AND METHODS

2.1. Sample Preparation
Ten samples of G. lucidum fruiting body and ten G. lucidum spores samples were originated from Taishan, China. The fruiting body of G. lucidum was cross-sectioned into thin slices. From each sample, multiple spectra were taken from five different positions: top surface, middle area, bottom surface, outer stipe and inner stipe. In total, 80 spectra from G. lucidum fruiting body samples and 30 spectra from G. lucidum spores samples were collected.

2.2. Spectral Acquisition
A Fourier transform infrared (FTIR) spectrometer (Perkin-Elmer Spectrum 100 model) with an attenuated total reflectance (ATR) accessory was used to record the absorbance spectra of the G. lucidum fruiting body and spores directly without any processing. The ATR-FTIR spectra of all the G. lucidum samples were recorded in the mid-IR region of 4000 - 400 cm^{-1} at resolution of 4 cm^{-1} with 20 scans for each spectrum. Each spectrum with high signal-to-noise signal of about 50 was obtained by an average of these 20 scans. Background spectra were always recorded before running the sample spectra in order to obtain absorbance spectra with smooth baseline and with minimum detection of absorption bands of water vapor and carbon dioxide present in the optical path of the infrared beam in the spectrometer. Strong absorption bands in the absorbance spectra of the G. lucidum samples were accurately obtained by applying sufficient pressure on the sample onto the ZnSe crystal using a diamond tip in the ATR set-up [5].

2.3. Spectral Pre-Treatment

ATR-FTIR spectra are affected by both the concentration of the chemical constituents and the physical properties of the analyzed product. The physical effects, such as baseline variation, light scattering, path length differences, etc., account for the majority of the variance among spectra while the variance due to chemical composition is considered to be small. Therefore mathematical pretreatments are essential to reduce the variation due to physical effects so as to enhance the contribution of the chemical composition [7] [8].

The spectra were first smoothed using the Savitzky-Golay algorithm [9] by spanning a 10-point window, and then were reduced by taking every sixth point to speed up subsequent manipulation. To remove the regions of the spectra with low signal-to-noise ratios arising from the lower system response, only the wavenumbers ranging from 4000 to 450 cm^{-1} with 593 spectra points were used in the analysis. To remove slope variation and to correct light scatter due to different particle sizes, the standard normal variate (SNV) method [10] was used to standardize each spectrum by setting its mean intensity to zero and the variance to one. After pretreatment, the mean spectra from fruiting body and spores of G. lucidum have similar patterns as shown in Figure 1, which indicates similar chemical composition. Some small between-class differences were observed in the region of 1800 - 1050 cm^{-1}, 3000 - 2700 cm^{-1} and 3700 - 3500 cm^{-1}, which may indicate different levels of the chemical components contained in fruiting body and spores of G. lucidum.

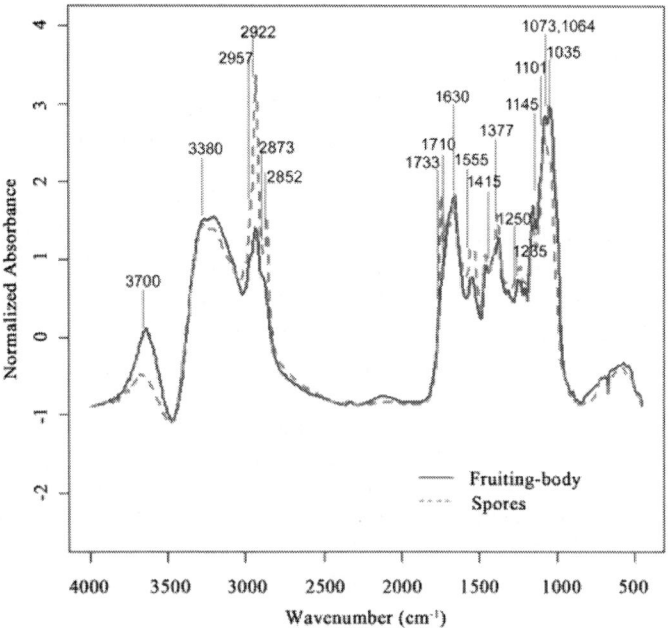

Figure 1. Mean spectra of fruiting body (solid line) and spores (dotted line) of G. lucidum after spectral pre-treatment.

2. MATERIALS AND METHODS

2.4. Statistical Analysis

1) Principal component discriminant analysis (PCDA) and partial least squares discriminant analysis (PLSDA)

Multivariate statistical methods including principal component analysis (PCA), partial least squares (PLS) and linear discriminant analysis (LDA) [11] - [13] were employed on whole spectral region to investigate the differences of spectra from spores and fruiting body of G. lucidum.

Simple and direct implementation of LDA in high-dimensional spectroscopic data setting provides poor classification results and the interpretation of the results is challenging due to singularity problem and highly-correlated spectral features. To solve this problem, many traditional approaches to this problem involve performing feature selection to reduce the variable dimension before classification. PCA and PLS were used to reduce the dimension of the original spectral data matrix with little loss of information. Then LDA focuses on finding a linear combination of the new variables, provided either by PCA or PLS, to construct canonical variate which best separates the two groups. Using pretreated spectral data described in Section 2.3, classification rules were derived using principal component discriminant analysis (PCDA) [14] - [16] and partial least squares discriminant analysis (PLSDA) [15] - [17]. The PCDA involves an initial PCA on the pre-treated spectra followed by a LDA performed on the first k PCs' scores. The PLSDA involves a PLS regression on the pre-treated spectra followed by a LDA on the first k PLS components' scores. Both PCDA and PLSDA were carried out with k ranging from 2 to 20.

2) Penalized discriminant analysis (PDA)

Though the spectroscopic data are highly correlated due to the presence of a large number of overlapping broad peaks, in many applications, only a small subset of variables (wavelengths/wavenumbers) contain sufficient information for discrimination. Hence there is interest in determining if a small subset of spectral wavenumbers containsas much information as the full spectrum does.

Penalized discriminant analysis (PDA) [18] [19] involves a sparse linear combination of all spectral features by imposing an L_1 norm penalty on the discriminant vector. The PDA extends LDA to the high-dimensional setting to deal with ill-posed problem without filtering out any spectral feature. Using pretreated spectral data with number of spectra points p = 593, classification rules were derived by a PDA using Fisher's criteria [20]. It generates discriminant vector v in Fisher's discriminant problem which involves a sparse linear combination of p number of spectral features by shrinking some coefficients towards zero and setting others equal to zero. The PDA investigated here can be expressed in the following optimization problem:

$$\max_v \left\{ v'\hat{\Sigma}_B v - \lambda \sum_{j=1}^{p} |s_j v_j| \right\} \text{ subject to } v'\hat{\Sigma}_w v \leq 1$$

where $\hat{\Sigma}_B$ is the sample estimate of between-class covariance matrix and $\hat{\Sigma}_w$ is the sample estimate of within- class covariance matrix, λ is non-negative tuning parameter. A diagonal estimate for the within-class covariance matrix is used here because it has been shown to give good results in the high dimensional setting [21] . Each feature of the discriminant vector v is additionally penalized by its sample within-group standard deviation s_j. The larger the value of λ, the larger the penalty on v and therefore the smaller the number of non-zero components, that is, the resulting discriminant vector v will be sparse in the spectral features. Classification accuracy was used to guide the choice of optimal parameter.

Leave-one-out cross-validation was used to train the algorithm by carrying out the PCDA, PLSDA and PDA classification rules on all the data except one sample which was then tested. This was repeated until all samples have been tested and an overall model accuracy was determined. To ensure that the wavelengths selected by PDA model are not training set specific, the PDA model was also validated with 70% of the data being treated as training data and 30% as test data. To ensure the statistical robustness, this process was repeated 50 times with different random splits of training and test sets, and the average misclassification rates were presented to assess the classification performance.

All the algorithms for computations and analyses were implemented in R statistical programming language [22] .

3. RESULTS AND DISCUSSION

3.1. Absorption Band Assignments of ATR-FTIR Spectra of G. lucidum

Figure 1 shows the typical ATR-FTIR spectrum of G. lucidum after pretreatment in the region of 4000 - 450 cm^{-1} with major peaks of the absorption bands labeled on the mean spectrum. The spectral wavenumbers and their corresponding assignments of the absorption bands in the ATR-FTIR spectrum of G. lucidum were given in Table 1 based on literature [7] [23] - [26] .

Polysaccharide, triterpene, sterols, amino acids, proteins, fatty acids have been known as the most biologically active substances in G. lucidum spores [27] [28] . The bioactive polysaccharides in the forms of glucomannan and arabinan isolated from G. lucidum spores, identified by the absorption band at 1064 cm^{-1}, 1035 cm^{-1} respectively (listed in Table 1), have been demonstrated to exhibit strong immunomodulation and anti-tumor activities including preventing oncogenesis and tumor metastasis [23] [24] . Furthermore, synergistic effect of polysaccharides and other bioactive components such as triterpene and sterols compounds isolated from spores G. lucidum, identified in the absorption band at 1415 cm^{-1}, 1377 cm^{-1} and 1145 cm^{-1} (given in Table 1), have shown to possess high bioactivities and proved effective as cytotoxic, antiviral and antiinflamatory agents [25] [27] . In the regions like ~1630 cm^{-1}, 1733 - 1710 cm^{-1}, 2957 - 2852 cm^{-1}, spectra receive contributions from the compounds, such as protein, fatty acids and lipids of G. lucidum spores [25] [27] .

3.2. Discrimination by PCDA and PLSDA

For PCDA and PLSDA model using the full spectrum region, the number of PCs or PLS components chosen is crucial to the discrimination performance. The discrimination results of cross-validation were used to optimize the number of PCs or PLS components. For PCDA model, the first seven PCs were used to construct the discrimination model and the leave-one-out cross-validation analysis gave a discrimination accuracy of 97%. With the relationship between the spectra variables and the responses taken into account for latent variable design, the PLSDA model used a fewer optimal number of latent variables (only three PLS components) when constructing the canonical variate and the leave-one-out cross-validation achieved a discrimination accuracy of 99%. In Figure 2 the 3D scatter plot of the first three PLS components illustrated a very clear separation between spores and fruiting body of G. lucidum in a 3D space.

Table 1. Absorption band assignments [7] [23] - [26] of the ATR-FTIR spectrum of G. lucidum.

Wave number (cm^{-1})	Functional Group Assignments
3500 - 3700	O-H stretching vibration of hydroxyl groups (mainly lipids and proteins)
3380	O-H stretching, Amine (N-H stretching vibration) mainly carbohydrates proteins
2957	CH_3 asym stretching (mainly lipids)
2922	CH_2 asym stretching (mainly lipids)
2873	CH_3 sym stretching (mainly proteins)
2852	CH_2 sym stretching (mainly lipids)
1710, 1733	C=O carbonyl stretching of saturated aliphatic esters
1630	Amide I (protein C=O stretch)
1555	Amide II (C-N, N-H stretching) mainly proteins
1415	O-H bending, polysaccarides
1377	Symmetric bending of aliphatic CH_3, triterpene compounds (CH_2=CH-CH_3)
1250	Pectic substances
1235	Amide III (C-N, N-H stretching) mainly proteins
1145	Cellulose (b-glucan), triterpene compounds (C-O)
1101	Antisym in-phase, pectic substances
1073	Rhamnogalactorunan, b-galactan
1064	C-O stretching, cell wall polysaccharides (glucomannan)
1035	OH and C-OH stretching in sugars, cell wall polysaccharides (arabinan)

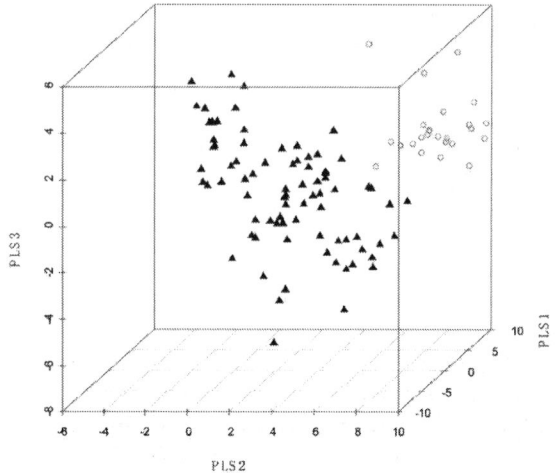

Figure 2. The 3D scatter plot of the first three PLS components' scores of the spectra from G. lucidum spores (○) and its fruiting body (▲).

3.3. Discrimination by PDA

PDA model seeks to find an optimal parameter λ with lowest error rate. Figure 3 shows the relationship among the tuning parameter λ, the number of the selected non-zero features (representing the sparsity level of the discriminant vector) and the corresponding error rate of PDA model. It illustrates that the sparsity level of the solution varies smoothly with λ, with larger values of λ resulting in very sparse solutions. When is less than 0.11, larger values of λ result in sparser solutions and lower error rate. The drop is observed at around 50 non-zero features. When λ is more than 0.11, larger λ gives higher error rate. The penalized LDA model with optimal tuning parameters of $\lambda = 0.11$ and 53 (out of 593) selected wavelengths points gives classification accuracy of 99% for discrimination between spectra of spores and fruiting body of G. lucidum. The contributions of this wavelength selection method are essential. It leads to a model having a small number of selected wavelength regions, yet comparable or better classification ability than full-spectrum model. Out of 50 random splits of data, the number of times the wavelength was selected shows how important is the wavelength to the discrimination. The selected wavelengths recorded inFigure 4 are concentrated in the regions of 3700 - 3500 cm^{-1}, 3000 - 2800 cm^{-1}, ~1700 cm^{-1}, ~1400 cm^{-1}, ~1000 cm^{-1}, which are consistent with the spectral variation in Figure 1. The selected wavelengths show us which parts of the spectrum are important to the discrimination.

It is possible that the superior performance of PDA model relative to the models using full wavelength is due to the fact that ATR-FTIR spectroscopic data consist of many overlapping absorption bands of which only a small proportion may be informative for explaining the response. Including those uninformative wavelength points in a model may introduce a great deal of noise and thus reduce the performance of the model.

3. RESULTS AND DISCUSSION

The good discrimination results from all these models suggested that there may exist some inherent compositional differences between spores and fruiting body of G. lucidum.

3.4. Correlation between Spectral Absorption Bands and Chemical Components of G. lucidum and Its Medicine Effect

Discrimination performance of the models may be explained by the correlation between spectral features and chemical constituents of G. lucidum spores. With non-zero features selected for discrimination, the discriminant vector of PDA model makes direct and valuable contribution to interpreting spectral features related to the medical fruiting body of G. lucidum. The non-zero regions with prominent absorption features around 1150 - 1050 cm^{-1} are typical features of triterpenoids and polysaccharides due to the C-O vibrations, which are the major chemical constituents of G. lucidum. The other prominent absorption peak selected at around 1700 cm^{-1} is consistent with a C=O stretching vibration in carbonyl compounds which may be characterized by the presence of high content of terpenoids and protein in G. lucidum. It is reported that G. lucidum spores possess a much higher content of triterpenoids on a weight basis when compared to G. lucidum fruiting body [29]. The selected features with sharp peak at around 3000 - 2700 cm^{-1} are due to C-H stretching vibration. The non-zero regions of 3700 - 3500 cm^{-1} are characteristic of carbohydrates proteins due to the O-H stretching vibration as shown in Table 1. The carbohydrate content in fruiting body of G. lucidum is much higher than that in spores [27] - [29], which also explains the mean spectra differences between fruiting body and spores of G. lucidum in this region.

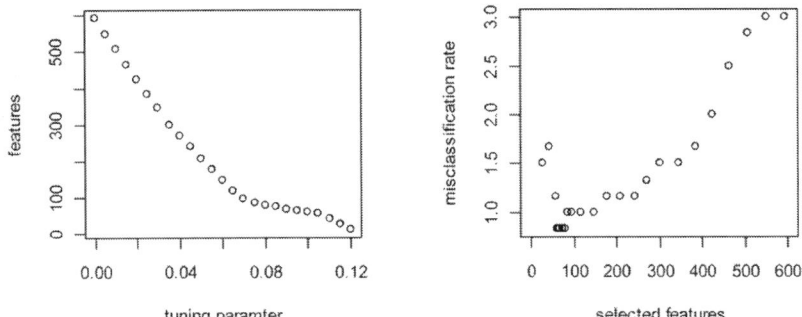

Figure 3. Left panel: Number of non-zero features versus tuning parameter; Right panel: Misclassification rate versus selected non-zero features.

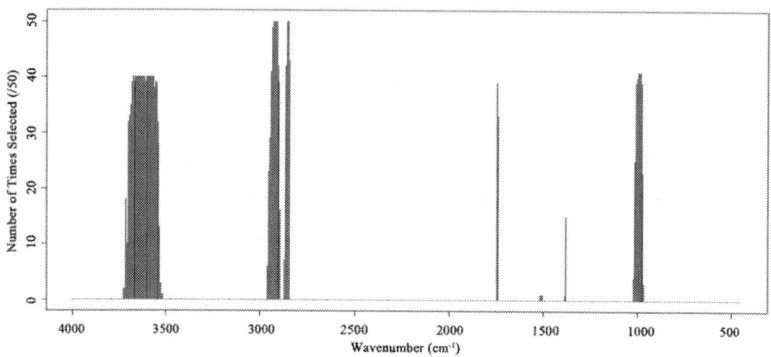

Figure 4. Wavenumbers selected for discrimination between spores and fruiting body of G. lucidum. The height of the bars shows the number of times the wavenumber was selected in 50 random splits of the data.

For the PCDA and PLSDA model, the PCDA or PLSDA loading of the original variables combines the loading from PCA or PLS and the loading from LDA when constructing a canonical variate. Therefore the PCDA loading and the PLSDA loading show the contribution at each wavelength to the linear diagnostic rule and thus can be related easily to the spectral features, which permits interpretation of its spectral basis. A comparison between PCDA loading and the PLSDA loading for discrimination between spores and fruiting body of G. lucidum can be seen in Figure 6. The loading features emphasized by the PLSDA model are very similar to those emphasized by PCDA model, which are consistent with group differences between spores and fruiting body of G. lucidum as shown in Figure 7. When comparing the PDA coefficient in Figure 5with PCDA and PLSDA loadings in Figure 6, we found that the non-zero regions with prominent absorption features from PDA model were also observed as the most prominent features by PCDA loadings and the PLSDA loadings. However, for the PLSDA and PCDA model, when the number of components used for the discrimination is getting bigger, the loadings from these two models may become more complex and thus the contribution of each wavelength to the classification becomes less interpretable when it is related to the spectral features. The major features of these loadings can also be explained by the assignments of the corresponding absorption bands in the ATR-FTIR spectrum listed in Table 1. The high consistency between the selected features by the discrimination models and chemical features of the ATR-FTIR spectrum may provide a quantitative explanation of the major chemical constituents of spores G. lucidum with respect to chemometrics.

3. RESULTS AND DISCUSSION

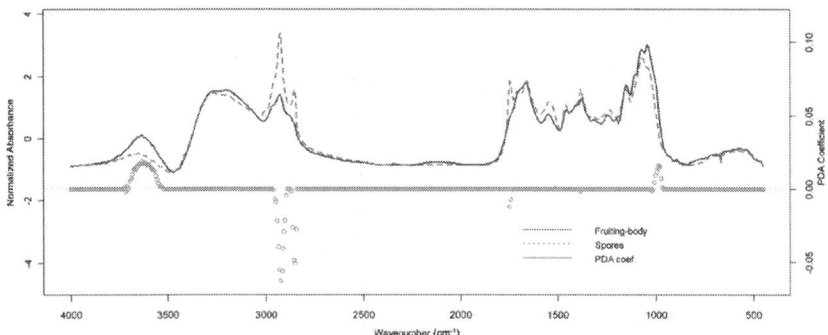

Figure 5. Mean spectra from G. lucidum spores (red line) and its fruiting body (blue line) with PDA coefficients in green for optimal tuning parameter.

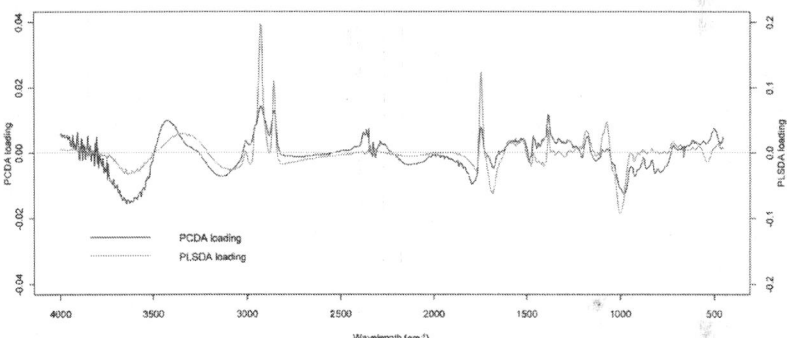

Figure 6. Comparison of PCDA loadings (blue line) and PLSDA loadings (green line) for discrimination between the spectra from fruiting body and spores of G. lucidum.

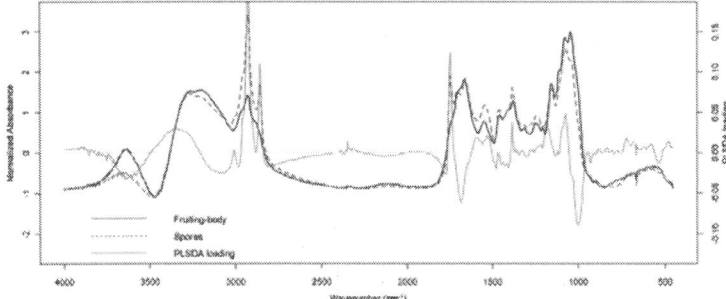

Figure 7. PLSDA loadings for discrimination between the spectra from spores and fruiting body of G. lucidum. The PLSDA loading is shown in green, with the mean spectra for the two types superimposed (fruiting body in blue line, spores in red line).

G. lucidum contains approximately 400 different bioactive compounds [22]. Among these ingredients, triterpenoids, polysaccharides, sterols, proteins, nucleo-sides and fatty acid are found the major chemical constituents of G. lucidum [28]. These biologically active compounds have been demonstrated to play a significant role in prevention of oncogenesis and tumor metastasis [6]. The spores of G. lucidum possess a much higher amount of bioactive substances than the fruiting body of G. lucidum [28]. In particular, polysaccharides from G. lucidum spores have immunomodulating properties. G. lucidum spores have been found to be effective in modulating the immune responses, and thus show efficacy of immunostimulatory and antitumor activities. Some comparative studies also reported that spores and fruiting body of G. lucidum showed different efficacy with regard to their antitumor effects and immunomodulatory activities [1] [29].

4. CONCLUSIONS

The spores of G. lucidum have higher economic value compared with its fruiting body due to the fact that numbers of bioactive substances of the spores are much higher than those of the fruiting body of G. lucidum. Since a variety of commercial G. lucidum products are available in various forms, it is of essential importance to distinguish spores and fruiting body of G. lucidum for the purpose of quality assurance.

In this study, the combination of ATR-FTIR spectroscopy and chemometrics method has been proved to be a very powerful tool to distinguish G. lucidum spores from its fruiting body efficiently. An excellent classification performance of up to 99% accuracy can be achieved by the discrimination models using the spectral features either selected or emphasized by the proposed models. By imposing penalties on the discriminant vectors, the PDA model presented in this paper enables an automatic selection of a small number of informative wavelength points to construct an efficacious discrimination model, which gives comparable or even higher accuracy than the PCDA and the PLSDA models based on the full wavelength.

Most essential contribution of the model is that the selected spectral regions for discriminant analysis show a good link between spectral features and chemical components of G. lucidum spores, which provided some evidence for its anticancer effect. This is a novel and important finding, as it provides quantitative interpretation and scientific support to the claims on the health benefits and antitumor properties of G. lucidum spores. It is also a potentially useful tool for quality control and fast discrimination of raw materials of traditional herbal medicine. Identified spectral regions may be targeted for further analysis linked with its active biochemical components of herbal medicine.

ACKNOWLEDGEMENTS

This research is supported by Academic Research Funds (AcRF: RI 12/10 TTL and AcRF: RI 6/14 ZY) of National Institute of Education, Nanyang Technological University, Singapore.

REFERENCES

1. Liu, X., Yuan, J.P. and Chen, X.J. (2002) Antitumor Activity of the Sporederm-Broken Germinating Spores of Ganodermalucidum. Cancer Letters, 182, 155-161.
2. Lin, Z.B. and Wang, P.Y. (2006) The Pharmacological Study of Ganoderma Spores. Journal of Peking University Health Science, 38, 541-547.
3. Chew, O.S., Hamdan, M.R., Ismail, Z. and Ahmad, M.N. (2004) Assessment of Herbal Medicines by Chemometrics-Assisted Interpretation of FTIR Spectra. Analytica Chimica Acta, 570, 116-123.
4. Lai, Y.H., Ni, Y.N. and Kokot, S. (2010) Classification of Raw and Roasted Semen Cassiae Samples with the Use of Fourier Transform Infrared Fingerprints and Least Squares Support Vector Machines. Applied Spectroscopy, 64, 649-656.
5. Stuart, B. (1997) Biological Applications of Infrared Spectroscopy. Analytical Chemistry of Open Learning (115). John Wiley & Sons, Chichester.
6. Dritsa, V. (2012) FT-IR Spectroscopy in Medicine. Infrared Spectroscopy-Life and Biomedical Science, 272-288.
7. Yap, K.Y.L., Chan, S.Y. and Lim, C.S. (2007) Authentication of Traditional Chinese Medicine Using Infrared Spectroscopy: Distinguishing between Ginseng and Its Morphological Fakes. Journal of Biomedical Science, 14, 265-273.
8. Woo, Y.A., Kim, H.J. and Cho, J. (1999) Identification of Herbal Medicines Using Pattern Recognition Techniques with Near-Infrared Reflectance Spectra. Microchemical Journal, 63, 61-70.
9. Savitzky, A. and Golay, M.J.E. (1964) Smoothing and Differentiation of Data by Simplified Least-Squares Procedures. Analytical Chemistry, 36, 1627-1639.
10. Barnes, R.J., Dhanoa, M.S. and Lister, S.J. (1989) Standard Normal Variate Transformation and De-Trending of Near-Infrared Diffuse Reflectance Spectra. Applied Spectroscopy, 43, 772-777.
11. Davies, A.M.C. and Fearn, T. (2005) Back to Basics: The Principles of Principal Component Analysis. Spectroscopy Europe, Tony Davies Column, 16, 20-23.
12. Næs, T., Isaksson, T., Fearn, T. and Davies, T. (2002) A User-Friendly Guide to Multivariate Calibration and Classification. NIR Publications, Chichester.
13. Barker, M. and Rayens, W. (2003) Partial Least Squares for Discrimination. Journal of Chemometrics, 17, 166-173.

14. Zhu, Y., Fearn, T., Samuel, D., Dhar, A., Hameed, O., Bown, S.G. and Lovat, L.B. (2008) Error Removal by Orthogonal Subtraction (EROS): A Customised Pre-Treatment for Spectroscopic Data. Journal of Chemometrics, 22, 130-134.
15. Roger, J.M., Palagos, B., Guillaume, S. and Bellon-Maurel, V. (2005) Discriminating from Highly Multivariate Data by Focal Eigen Function Discriminant Analysis; Application to NIR Spectra. Chemometrics and Intelligent Laboratory Systems, 79, 31-41.
16. Zhu, Y. and Tan, T.L. (2015) Discrimination of Wild-Grown and Cultivated Ganoderma lucidum by Fourier Transform Infrared Spectroscopy and Chemometric Methods. American Journal of Analytical Chemistry, 6, 480-491.
17. Lelong, C.C.D., Roger, J.M., Brégand, S., Dubertret, F., Lanore, M., Sitorus, N.A., Raharjo, D.A. and Caliman, J.P. (2010) Evaluation of Oil-Palm Fungal Disease Infestation with Canopy Hyperspectral Reflectance Data. Sensors, 10, 734-747.
18. Hastie, T., Buja, A. and Tibshirani, R. (1995) Penalized Discriminant Analysis. The Annals of Statistics, 23, 73-102.
19. Hastie, T., Tibshirani, R. and Friedman, J. (2009) The Elements of Statistical Learning: Data Mining, Inference and Prediction. Springer, New York.
20. Witten, D.M. and Tibshirani, R. (2011) Penalized Classification Using Fisher's Linear Discriminant. Journal of the Royal Statistical Society: Series B, 73, 753-772.
21. Dudoit, S., Fridlyand, J. and Speed, T. (2001) Comparison of Discrimination Methods for the Classification of Tumors Using Gene Expression Data. The Journal of the American Statistical Association, 96, 1115-1160.
22. R Core Team (2012) R: A Language and Environment for Statistical Computing. R Foundation for Statistical Computing, Vienna.
23. Wang, Y.L. (2012) Application of Fourier Transform Infrared Microspectroscopy (FTIR) and Thermogravimetric Analysis (TGA) for Quick Identification of Chinese Herb Solanum lyratum. Plant Omics Journal, 5, 508-513.
24. Jin, W.Y., Wan, C.Y. and Cheng, C.G. (2015) Study on the Identification of Radix Bupleuri from Its Unofficial Varieties Based on Discrete Wavelet Transformation Feature Extraction of ATR-FTIR Spectroscopy Combined with Probability Neural Network. International Journal of Analytical Chemistry, 2015, Article ID: 950209.

REFERENCES

25. Chen, X.L., Liu, X.C., Sheng, D.P., Huang, D.K., Li, W.Z. and Wang, X. (2012) Distinction of Broken Cellular Wall Ganoderma lucidum Spores and G. lucidum Spores Using FTIR Microspectroscopy. Spectrochimica Acta Part A, 97, 667-672.
26. Gorgulu, S.T., Dogan, M. and Severcan, F. (2007) The Characterization and Differentiation of Higher Plants by Fourier Transform Infrared Spectroscopy. Applied Spectroscopy, 61, 300-308.
27. Wang, X., Chen, X.L., Qi, Z.M., Liu, X.C., Li, W.Z. and Wang, S.Y. (2012) A Study of Ganoderma lucidum Spores by FTIR Microspectroscopy. Spectrochimica Acta Part A: Molecular and Biomolecular Spectroscopy, 91, 285-289
28. Russell, R. and Paterson, M. (2006) Ganoderma—A Therapeutic Fungal Biofactory. Phytochemistry, 67, 1985-2001.
29. Yue, G.L., Fung, K.P., Leung, P.C. and Lau, B.S. (2008) Comparative Studies on the Immunomodulatory and Antitumor Activities of the Different Parts of Fruiting Body of Ganoderma lucidum and Ganoderma spores. Phytotherapy Research, 2, 1282-1291.

CHAPTER 14

Chemometric Analysis of an Sanitary Landfill Leachate

Ana M. C. Grisa, Cíntia Paese, Oclide José Dotto, Ronaldo Nicola, Mara Zeni

Universidade de Caxias do Sul, Caxias do Sul, Brazil

ABSTRACT

This paper presents a study on the biotic/abiotic conditions of the São Giácomo sanitary landfill, located near the city of Caxias do Sul, Brazil, through statistical analysis of fourteen physic-chemical data sets for the leachate, produced in the garbage dump site over a long period of years. Different chemometric methods are used in the statistical analysis. For example, the correlations between the variables, related to the degraded organic matter and biological activity, are determined by means of multivariate methods. The results highlight that BOD, COD, VTS, FTS and TS give information on the anaerobic degradation of the organic matter contained in the cells, and suggest that the greater the contribution of the variables with positive weights in PC1 the greater the level of organic matter degradation. The variables TN, Amon Nit. and alkalinity are related to the biological activity and determine the potency of the variables in relation to time. The greater the contribution of the variables related to organic degradation the greater the values in PC2 and the lesser the potency of these variables, whose influence is greater in the second stage of anaerobic degradation. The variables of PC2 is important plans of the contamination of the leached in the bodies hídrics.

Keywords: Landfill Leachate; Physico-Chemical Variables; Chemometric Method; Principal Components Analysis

1. INTRODUCTION

A sanitary landfill is predominantly an anaerobic biological system, in which the treatment of the deposited domestic solid residues occurs through the interaction of different microbial species. The development of the biosistem allows the conversion of the organic matter into methane gas and percolated liquids. The percolated/lixiviated liquids, commonly refered to as leachate, are the result of the enzymatic action of the microorganisms of the system and the resulting products of biodegradation [1]. The microbial activity in each stage of the solid residue stabilization process gives typical character ristics to the leachate. The variation in the leachate quality is generally attributed to a myriad of interacting factors such as type and depth of the solid waste, age of fill, rate of water flow and the interaction of the leachate with its environment [2,3].

Leached him it is the main source poluidora of the underground waters and surfaces. The impact produced for the leached on the environment it is directly related with its phase of decomposition [4].

The stabilization degree, organic residue mineralizetion, organic matter distribution and the intervention of diverse groups of bacteria and fungi that act simultaneously influence, with regard to the leachate composition, the anaerobic digestion. In the anaerobic degradation, nitrogen is important for determining the nutrient sufficiency of the conditions (aerobic and anaerobic) at each stage of the stabilization [5,6].

Environmental concerns regarding methanogenic-state landfill leachates, characterized by low organic carbon concentration, high ammonia-N levels, and a large spectrum of metals. The slow leaching of nitrogen from solid waste in landfills, resulting in high concentrations of ammonia in the landfill leachate, may last for several decades [7].

The main potential effects of the release of the leached in bodies hidric associated to the decrease of the concentration of the dissolved oxygen, to the toxic caused by the NH_3 and the eutrophication due to the high concentrations of N_2 [8].

The area named Vazadouro de São Giácomo, in the peripheral region of Caxias do Sul, RS, Brazil, on the banks of the Tega River, received in a disorderly form, for a period 1988 to 1990, the urban solid residues produced in the city, causing the environmental degradation of an area of approximately 1.4 hectares (**Figure 1**).

The remediation project for the area degraded by solid waste was prioritized aiming to minimize the generation of percolated liquids and to carry out environment monitoring. The arrangement of the solid waste in layers or cells was executed according to the NBR 8418/1984 and the NBR 8419/1984 norms, and the resulting system is monitored monthly.

The most easily found post-consumption thermoplastic polymers, defined as commodities, are: low density polyethylene (LDPE), high density polyethylene (HDPE), polypropylene (PP), polyvinyl chloride (PVC) and polystyrene (PS). It has been estimated that in Europe the consumption of thermoplastics is 80 kg/habitant/year and in Brazil it is around 23 kg/habitant/year [9]. Since these polymers have a high resistance to degradation, large volumes are discarded to garbage dump sites and industrial waste sites.

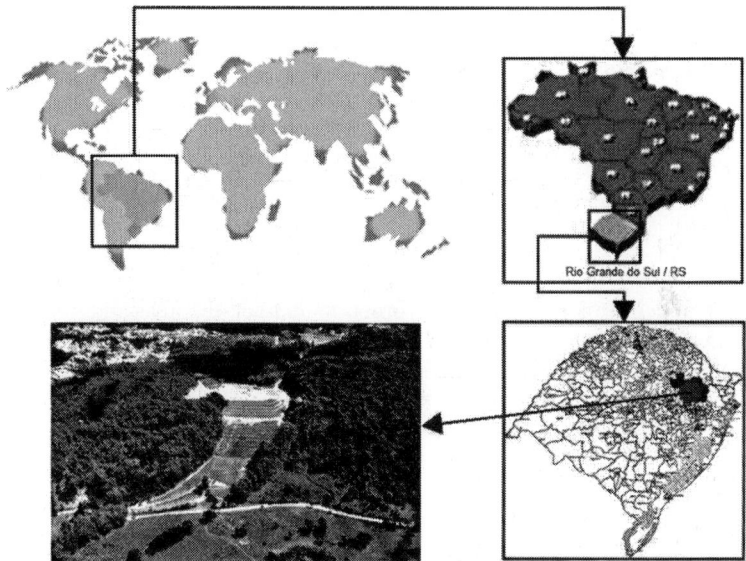

Figure 1. The sanitary landfill São Giácomo near the city Caxias do Sul, RS, Brazil.

In this study fourteen physic-chemical variables of the leachate are analyzed along with the factors which influence the organic matter degradation at the São Giácomo sanitary landfill, consisting of seven refuse cells of different ages, by means of principal components analysis (PCA). In this work the objective was to study the degradation of organic substances and the biological activity of the physic-chemical leachate variables in the cells of different ages from the sanitary landfill.

Chemometric analysis method and principal component analysis were used for interpretation and to indicate associations between the variables.

Later on, in another work, these factors will be considered in relation to the biodegradation or oxidative degradation of the organic matter and the buried polymers in the landfill.

2. EXPERIMENTAL PART

2.1. Physico-Chemical Parameters

The data set for fourteen physico-chemical parameters was compiled over ten years by the Instituto de Saneamento Ambiental (Environmental Health Institute of the University of Caxias do Sul) and by the Companhia do Desenvolvimento (Development Company of Caxias do Sul) (CODECA) [10], and relates to leachate flowing from the São Giácomo sanitary landfill. This landfill comprises seven cells of different ages. The cells are denoted as follows (the numbers inside parentheses indicate the age in months): C1 (136), C3 (115), C4 (94), C5 (54), C7 (43), C8 (27) and C9 (20).

The samples were analysed with respect to fourteen parameters: pH, alkalinity, total solids (TS), volatile total solids (VTS), fixed total solids (FTS), chemical oxygen demand (COD), biochemical oxygen demand (BOD), total nitrogen (TN.), amoniac nitrogen (Amon Nit.), cadmium (Cd), copper (Cu), iron (Fe), manganese (Mn) and zinc (Zn). The chemical components to be analyzed were chosen on a list of priority degradation and biological activity in the organic matter.

The variables collected at the sanitary landfill (**Figure 1**) were correlated and analyzed by chemometric methods using multivariate statistics in the treatment and chemical analysis in order to extract the maximum chemical information. The chemometric methods were applied through the classification and comparison of different samples [11] by determining correlations between the degradation of organic substances and the biological activity.

2.2. Multivariate Analysis

The physic-chemical variables of the leachate were investigated by different chemiometric methods: basic statistical methods for the determination of mean and values, standard deviations, minimum and maximum, and the principal components analysis method for advanced information, everything through the software Statistical Product and Service Solution (SPSS).

The problems originated from some missing values were minimized using the procedure "exclude marry listwise" available in SPSS. The variables were standardized to remove the influence of the measures units and the values magnitudes.

Before using PCA, mathematical manipulation of the data set was carried out, describing, investigating and comparing the variables in the set, aiming at predicting models through the multivariate analysis (median, mean values, standard deviations, etc.).

The multivariate analysis was performed through PCA [12-14], which is an efficient way to suppress redundant information on the multidimensional initial data and to obtain the relevant non-correlated information. This means that the covariance matrix of the variables $X_1, X_2, ..., X_N$ containing the data is diagonal or quasi diagonal.

Initially we have a matrix $[X_1, X_2 ... X_N]$ of the data which is transformed to the centered form (mean-deviation form), or to another convenient form for analysis. Let us assume that the example matrix shown is already in such a form. The aim of PCA is to find an orthogonal matrix $P = [u_1 \ u_2 ... \ u_p]$, that is, a matrix P such that $P^t P = I$, which defines a change of variable $X = PY$; with another notation,

$$\begin{bmatrix} x_1 \\ x_2 \\ \vdots \\ x_p \end{bmatrix} = \begin{bmatrix} \mathbf{u}_1 & \mathbf{u}_2 & \cdots & \mathbf{u}_p \end{bmatrix} \begin{bmatrix} y_1 \\ y_2 \\ \vdots \\ y_p \end{bmatrix},$$

where the new coordinates y_1, y_2, ... , y_3 are uncorrelated and are displayed in descending order of variance. The unitary vectors u_1, u_2, ..., u_p, arranged in ascending order of importance, are named Principal Components (PCs) of the data (observations matrix). These PCs are described in terms of eigenvalues of some matrix.

The first principal component (PC1) u_1 determines a new variable y_1 in the following way. Let c_1, c_2, ..., c_p be the u_1 coordinates. Since the transposed vector $\mathbf{u}_1^t$ is the first line of P^t, the equation $Y = P^t X$ show that,

$$y_1 = \mathbf{u}_1^t \mathbf{X} = c_1 x_1 + c_2 x_2 + \cdots + c_p x_p \tag{1}$$

We see that y_1 is a linear combination of the original variables x_1, x_2, ..., x_p, that have the coordinates of u_1 as its coefficients. In the same way, u_2 determines a variable y_2, ecc.

So, by means of the PCA we obtain the least number of uncorrelated variables, which are a linear combination of the initial ones, allowing the description of the structure and the interconnections of the original variables of the phenomenon being studied, with base on the PCs obtained. The results of the PCA can be visualized by graphs of the scores which allow an estimation of the influence of each variable of each sample [15,16].

The representation and analysis of weights and scores graphs obtained from the relations between the physicchemical variables of each sanitary landfill cell, per lifetime, determines the probability distribution. Through the interpretation of the resulting data, correlating these data and that for the other cells of the landfill and using PCA, the factors which account for the maximum variance in all observed variables are obtained. The choice of PCs to describe the data was made considering the percentage of the variance they describe and the residual variance.

3. RESULTS AND DISCUSSION

After determining the mean value and standard deviation, the correlation was studied for all measured varaibles. The data were autoscaled (mean equal zero and variance equal 1). The application of the main components is to reduce the dimensions of the original variables and for the extraction of the main components it is selected the components main through of the scree plots.

The sedimentation graphs (eigenvalue relations and component numbers) obtained from the PCA indicate the number of PCs that were used for each cell. The adopted approach refers the use of the eigenvalues that they possess larger capacity to explain the larger total variability. The percentages of variances in resulting eigenvectors (PCs) [16-18].

An analysis of the correlation matrix allowed to uncover strong associations between some variables PCA showed the existence of the two a at four a significant PCs.

It is reasonable to expect that two to four main components were enough to describe the original variables of each cell of the sanitary landfill. In fact, according **Table 1**, two PCs for C5 and C7 are sufficient to describe the data, since they contain 79.93% and 80.52%, respectively, of the original results, while, for the cell C1, four PCs contain 77.75% of the variability of the data. The results for cell C1 demand four PCs due to the disorderly arrangement of the solid waste in the landfill, which requires a larger number of representative variables. For cells C3, C4, C8 and C9 three PCs are needed to describe the original data, since they contain 69.20%, 81.55%, 88.911% and 85.27%, respectively, of the original results of the data.

It can be observed that for the more recent cells the variance percentage of the original data is greater, due to the high level of organic matter degradation (major activity). The other cells show signs of stabilization, since their degradation stages occur under neutral and alkaline pH conditions.

In **Table 1**, PCi, i = 1, 2, 3, 4, indicates the ith principal component. As it can be seen in the table, cells C1 (39.95%), C4 (53.52%), C5 (50.11%), C7 (53.50%) and C8 (54.17%) give the variables COD, BOD, VTS, TS, FTS as corresponding to first principal component (PC1). Except for C3 and C9, the cells have as PC1 the correspondent variables BOD, COD, VTS, FTS and TS, indicative of anaerobic degradation of the organic matter in the cells.

The second principal component (PC2), **Table 1**, for cells C1, C3, C4, C5, C7, C8 and C9, are associated to the variables TN., Nit. Amon. and alkalinity, that control the acidity in the development of the bacterial biological activity.

Due to the low percentage of metals, the analysis for PC3 was not performed in this study. Even so, the presence of metals should be considered in the landfill, since they are responsible for movement and complexation in the formation of acids, and complexation and precipitation of the metallic species during methanogenic fermentation.

The PCA results are visualized in graphs that identify clusters of physico-chemical variables values. The weights graphs (Figures 2(a) and 3(a)) contain important information that allows us to describe the variation in the original data, and the scores graphs (Figures 2(b) and 3(b)) exhibit information about the age of the cells, making similarities, clusters and differences visible, based on the used variables.

From the behavior of the physico-chemical variables in terms of lifetime (Figures 2(a) and 3(a)) it can be concluded that the greater the contribution of the variables with positive weights in PC1 the greater the degradation of the organic matter (variables in PC1). The negative values for the weights in PC2 demonstrate that the variation in the values for the physico-chemical variables influences the second stage of anaerobic degradation.

On considering cells C5 and C7 (Figures 2(a) and (b)), with respective ages of 54 and 42 months, in the scores graph (Figures 2(b)) for PC1, the group on

3. RESULTS AND DISCUSSION

the left side decreases over time, due to the degradation of organic matter consumed in the landfill.

PC2 is related to the biological activity which increases over time. The variables (more active variables) in PC2, located on the lower part of the scores graph (Fig ures 2(b)), are more meaningful (i.e. potent) in relation to the biological activity, while the less potent ones have positive scores.

Table 1. Percentage of variance and PCs (physic-chemical parameters) for the São Giácomo sanitary landfill.

Cell	CPs	% Variance	CPs variables
C1	PC1	39.954	BOD, COD, VTS, FTS, TS, alcalinity, Zn
	PC2	16.097	BOD, FTS, alkalinity, Amon. Nit., Total Nit, pH, Cd
	PC3	13.456	Cu, Mn, -pH
	PC4	8.249	Cd, Mn, Fe
C3	PC1	33.54	FTS, TS, alkalinity, Amon. Nit., Total Nit., pH
	PC2	23.037	BOD, COD, VTS, TS, -Amon.Nit., Cu, Zn, -Cd
	PC3	12.627	Total Nit., -Cu, -Zn, Mn, Fe
C4	PC1	53.527	BOD, COD, FTS, VTS, ST, Cd, Zn, Cu, -pH, Nit. Amon.
	PC2	20.584	alkalinity, -Amon. Nit., Total Nit., -Cd
	PC3	7.446	Fe, Mn
C5	PC1	50.113	BOD, COD, FTS, STV, ST, Fe, Zn, Mn, -pH
	PC2	29.283	FTS, alkalinity, Amon. Nit., Total Nit., Cu, -Cd
C7	PC1	53.507	BOD, CDO, FTS, VTS, TS, Mn, Zn, Cu, Fe, -pH
	PC2	27.018	FTS, alkalinity, Amon. Nit., Total Nit., -Fe, pH
C8	PC1	54.172	BDO, COD, FTS, VTS, ST, Fe, Mn, Zn, Cu, -pH
	PC2	26.938	FTS, alkalinity, Amon. Nit., Total Nit, pH
	PC3	7.801	Zn, Cd
C9	PC1	54.381	BOD, COD, -alkalinity, -Total Nit., Fe, Mn, -pH FTS, VTS, -Amon. Nit. Amon., Zn
	PC2	21.511	VTS, -FTS, -Amon.Nit., Zn
	PC3	9.378	Cu, Cd, -Zn
C10	PC1	79.838	BOD, COD, FTS, VTS, ST, Amon. Nit., Total Nit., Fe, Mn, alkalinity, Zn
	PC2	18.877	FTS, Amon. Nit., Total Nit., -Zn, pH

The organic matter degradation is aided by the presence of nitrogen, which is important in the determination of the nutrient sufficiency of the aerobic and anaerobic conditions during each stabilization phase.

For cells C3, C4, C8 and C9, with 115, 94, 27 and 20 months of age, respectively, three PCs were considered and for cell C1, with 136 months of age, four PC's were considered.

In order to visualize the weights and scores of cells C1, C3, C4, C8 and C9, aiming a better interpretation, twodimensional graphs of PC1 versus PC2 and PC1 versus PC3 (Figures 3(a) and (b)) were constructed. The interpretation of the two-dimensional scores graph (**Figure 3**(b)) allows a prediction, based on PC2, of the most significant time values in relation to the biological activeity on the organic matter (potency).

Since cell C1 has a disorderly arrangement of residues, PC2 (**Figure 3**(b)) has higher values in the scores graph for the numeric variables at the upper part and less potent ones in the lower part. It may therefore be concluded that degradation of the organic matter takes place in the final months and that, for the variables with numeric values from 40 to 85, the biological activity reflects the more potent variables. In this cell, the degradation of the organic matter occurs more intensely over time, but one can note a less potent biological activity.

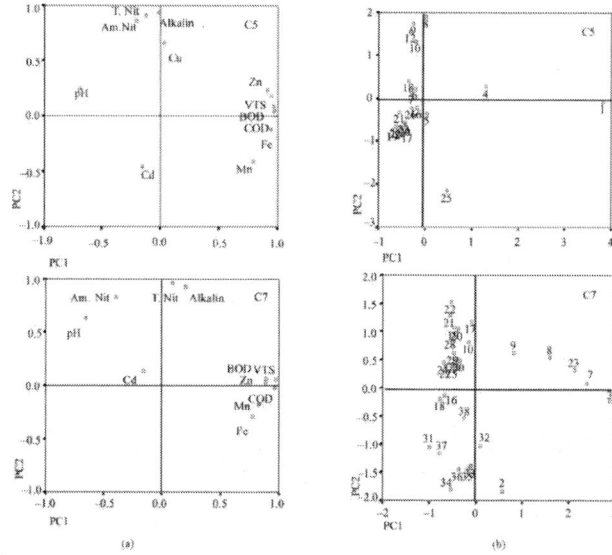

Figure 2. Graph of (a) weight (b) scores for cells C5 and C7.

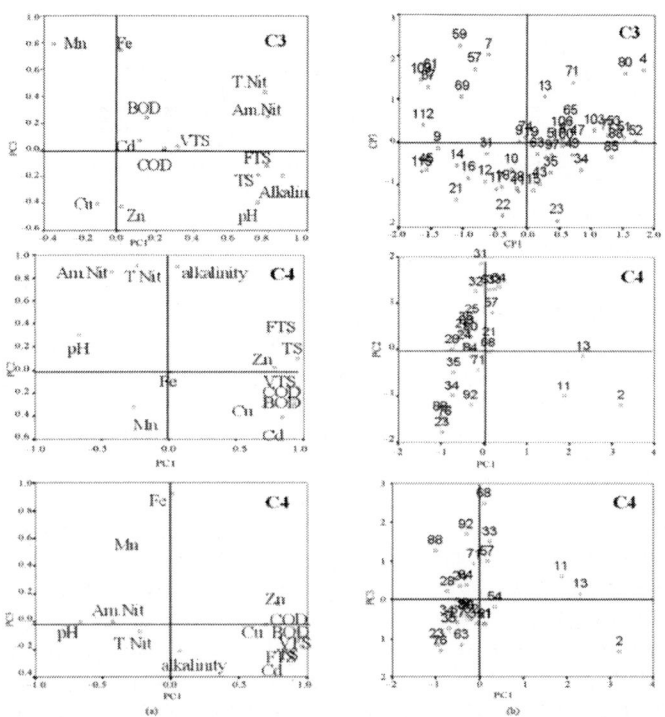

3. RESULTS AND DISCUSSION

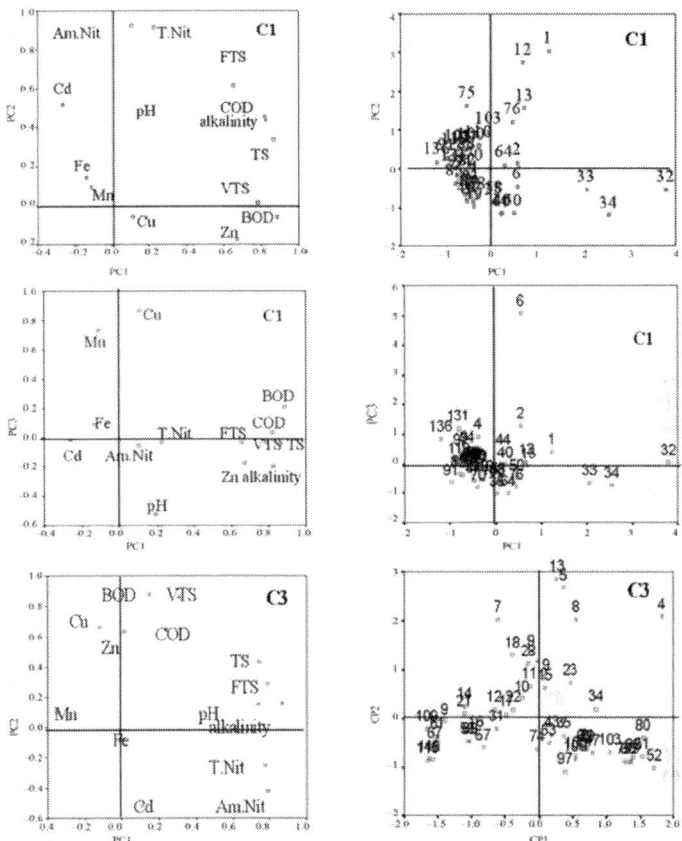

Figure 3. Physico-chemical parameters PC1 versus PC2 and PC1 versus PC3 (a) weight (b) scores for cells C1, C3 and C4.

Considering cells C4 and C8, for PC2, the most potent variables are located on the lower part of the scores graph (**Figure 3**(b)), while the less potent ones are on the upper part. These positions in the weights graph (Figures 3(a)), PC1 versus PC2, indicate that the greater the contribution of the variables related to organic degradation the greater the score values in PC2, and the lesser the potency of these variables.

From an analysis of the newest cell C9, it can be seen that its potent variables are on the lower and the upper parts of the graph, indicating that the degradation of organic matter and the biological activity occur simultaneously

An analysis of the biodegradability of the organic matter is of fundamental importance in the study of the abiotic factors that act on the sanitary landfill ecosystem. Such an analysis can reveal the presence of anaerobic microbial activity, considering that the leachate of cells C1, C3 and C4 has predominantly methanogenic characteristics (in 70% of the cases the BOD/COD ratio was below 0.4).

The newest cells C5, C6, C7 and C9 are in an unstable methanogenic phase (BOD/COD > 0.4), and an overlap of the stabilization stages can be observed during the degradation process, due to the heterogeneity of the material deposited in the landfill.

4. CONCLUSIONS

The physic-chemical and biological processes in a sanitary landfill, which lead to the degradation of the organic components of the dumped mass of solid waste, can be monitored through the collection of data that served to indicate the decomposition state of the dumped mass and the transformations that go on at the site. The sanitary landfill can be considered as a biological reactor, where there are suitable conditions for the growth of the bacteria that are responsible for biodegradation.

The chemometric analysis of the leachate data coming from São Giácomo sanitary landfill in Caxias do Sul, according the principal components analysis of fourteen physico-chemical parameters obtained from the leachate relating to seven studied cells, showed that two to four PCs are enough to describe the original data and identified the physic-chemical parameters related to the anaerobic digestion (enzymatic action) in terms of the leachate composition [19,20].

It could be observed that for the newest cells, the percentage of variance in the original data is greater than for others, due to the high concentration of organic matter undergoing degradation (more activity), the older cells showing signs of stabilization.

The PCA results highlight, in PC1 (BOD and COD, VTS, FTS and TS), variables that give information on the anaerobic degradation of the organic matter contained in the cells, and suggest that the greater the contribution of the variables with positive weights in PC1 the greater the level of organic matter degradation. The PC2 variables (TN., Amon Nit. and alkalinity) are related to the biological activity and determine the potency of the variables in relation to time. The greater the contribution of the variables related to organic degradation the greater the values in PC2 and the lesser the potency of these variables, whose influence is greater in the second stage of anaerobic degradation.

A sanitary landfill does not have a single age, but a variation of associated ages. These complex ecosystems converge toward the final stability of the organic matter, after countless transformations and biochemical interact tions between different microbial species, that can be modeled by means of chemometric analysis through the study of the physic-chemical parameters of the sanitary landfill. The method of multivariate analysis of the data (PCA) proved to be efficient to analyze and to characterize the leachate of the sanitary landfill.

The PC2 variables (TN., Amon Nit. and alkalinity) are related to the biological activity e quase totalidade do Amon Nit. in leached him it is originating from of the degradation of the matter orgânica. The effluent pouring out with significant amounts of nitrogen in a receiving body it can cause decomposition, decrease of the concentration of dissolved oxygen and toxicidade to the environment. It is recommended to accomplish treatment of removal of the ammonia from leachate before it discards it in the environment.

ACKNOWLEDGEMENTS

The authors thank the CNPQ, FAPERGS and University Caxias do Sul for financial support.

REFERENCES

1. S. R. Qasin and W. Chiang, "Sanitary Landfill Leachate: Generation, Control and Treatment," Technomic Publishing Company, Lancaster, 1994.
2. G. Andreotta and P. Cannas, "Chemical and Biological Characteristics of Landfill Leachte," In: Landfilling of Waste: Leachate, 2nd Edition, Chapman and Hall Ltd, London, 1997, pp. 65-88.
3. D. Augenstein and J. Pacey, "Modeling Landfill Methane Generation," Proceedings of 3rd International Landfill Symposium, Cagliari, 1991, pp. 115-148.
4. L. F. Sá, "Evaporação Natural do Lixiviado do Aterro de Muribeca Através de um Destilado Solar. Master, Programa de Pós Graduação em Engenharia Civil, Centro Tecnologia e Geociências, Universidade do Pernambuco (UFPE), Recife, 2008.
5. T. H. Christensen and R. Cossu, "Landfill Leachate: An Introduction," In: Landfilling of Waste: Leachate, 2nd Edition, Chapman and Hall Ltd, London, 1997, pp. 3-13.
6. J. M. Lema, R. Mendez and R. Blazquez, "Characteristics of Landfill Leachates and Alternatives for Their Treatment: A Review," Water, Air and Soil Pollution, Vol. 40, No. 3-4, 1988, pp. 223-250.
7. P. Y. Jokela, R. H. Kettunen, K. M. Sormunen and J. A. Rintala, "Biological Nitrogen Removal from Municipal Landfill Leachate: Low-Cost Nitrification in Biofilters and Laboratory Scale in-Situ Denitrification," Water Research, Vol. 36, No. 16, 2002, pp. 4079-4087.
8. P. Kjeldsen, M. A. Barlaz, A. P. Rooker, A. Bauna and T. H. Christensen, "Present and Long-Term Composition of MSW Landfill Leachate: A Review," Critical Reviews in Environmental Science and Technology, Vol. 32, No. 4, 2002, pp. 297-336.
9. M. Avella, E. Bonadies, E. Martuscelli and M. Rimedio, "European Current Standardization for Plastics Packaging Recoverable though Composting and Biodegradation," Polymer Testing, Vol. 20, No. 5, 2001, pp. 517-521.

10. CODECA, Companhia de Desenvolvimento de Caxias do Sul Relatório de Monitoramento Ambiental do Aterro Sanitário São Giácomo, Caxias do Sul, 2001.
11. D. B. Voncina, D. Dobcnik, M. Novic and J. Zupan, "Chemometrics Characterization of the Quality of River Water," Analytical Chimica Acta, Vol. 462, No. 1, 2002, pp. 87-100.
12. B. Walczak and D. L. Massart, "Dealing with Missing Data," Chemometrics and Intelligent Laboratory Systems, Vol. 58, No. 1, 2001, pp. 15-27.
13. M. P. Kallio, S.Mujunen, G. Hatzimihalis, P. Koutoufides, P. Minkkinen, P. Wilkie and M. Connor, "Multivariate Data Analysis of Key Polluants in Sewage Samples: A Case Study," Analytica Chimica Acta, Vol. 393, No. 1-3, 1999, pp. 181-191.
14. I. T. Jolliffe, "Principal Component Analysis," SpringerVerlag, New York, 2002.
15. K. Singh, A. Malik, V. K. Singh, D. Mohan and S. Sinha, "Chemometrics Analysis of Groundwater Data of Alluvial Aquifer of Gangetic Plain, North India," Analytica Chimica Acta, Vol. 550, No. 1-2, 2005, pp. 82-91.
16. M. Sena and R. Poppi, "Avaliação do uso de Métodos Quimiométricos em Análise de Solos," Química Nova, Vol. 23, No. 2000, pp. 547-556.
17. C. Pérez, "Técnicas Estadísticas com SPSS," Pearson Educacion, Madrid, 2001.
18. J. E. Jackson, "A User Guide to Principal Components," John Wiley & Sons, Inc., Hoboken, 2003.
19. A. M. C. Grisa and M. Zeni, "Estudio de la Degradación de los Polímeros Commodities en el Medio Ambiente," Tese de Doutorado, León, 2004.
20. D. Kulikowska and E. Klimiuk, "The Effect of Landfill Age on Municipal Leachate Composition," Bioresource Technology, Vol. 99, No. 13, 2008, pp. 5981-5985.

CHAPTER 15
Profiling of Fatty Acids Composition in Suet Oil Based on GC-EI-qMS and Chemometrics Analysis

Jun Jiang [1,2] *and Xiaobin Jia* [1,2,*]

[1]Affiliated Hospital on Integration of Chinese and Western Medicine, Nanjing University of Chinese Medicine, Xianlin Avenue 138#, Xianlin University City, Nanjing 210023, China
[2]Key Laboratory of New Drug Delivery System of Chinese Meteria Medica, Jiangsu Provincial Academy of Chinese Medicine, 100# Shizi Road, Nanjing 210028, China

ABSTRACT

Fatty acid (FA) composition of suet oil (SO) was measured by precolumn methylesterification (PME) optimized using a Box–Behnken design (BBD) and gas chromatography/electron ionization-quadrupole mass spectrometry (GC–EI-qMS). A spectral library (NIST 08) and standard compounds were used to identify FAs in SO representing 90.89% of the total peak area. The ten most abundant FAs were derivatized into FA methyl esters (FAMEs) and quantified by GC–EI-qMS; the correlation coefficient of each FAME was 0.999 and the lowest concentration quantified was 0.01 µg/mL. The range of recovery of the FAMEs was 82.1%–98.7% (relative standard deviation 2.2%–6.8%). The limits of quantification (LOQ) were 1.25–5.95 µg/L. The number of carbon atoms in the FAs identified ranged from 12 to 20; hexadecanoic and octadecanoic acids were the most abundant. Eighteen samples of SO purchased from Qinghai, Anhui and Jiangsu provinces of China were categorized into three groups by principal component analysis (PCA) according to the contents of the most abundant FAs. The results showed SOs samples were rich in FAs with significantly different profiles from different origins. The method described here can be used for quality control and SO differentiation on the basis of the FA profile.

Keywords: Suet oil (SO); Composition profiles; Fatty acids (FAs); GC–EI-qMS; Chemometrics analysis

1. INTRODUCTION

Suet oil (SO), a fatty oil obtained from the domestic goat (*Capra hircus* Linnaeus) or sheep (*Ovis aries* Linnaeus), has been used in the food industry [1] and the medicine industry [2]. SO is rich in unsaturated and saturated fatty acids (FAs) [3], which are involved in a number of important physiological processes. They provide energy to the cell and act as substrates in the synthesis of fats, lipoproteins, liposaccharides and eicosanoids [4]. Furthermore, SO can be used as an excipient for enhancing the efficacy of traditional Chinese medicines such as Epimedum (Berberidaceae). It was hypothesized that the beneficial effects of Epimedium could be attributed to promotion of the intestinal absorption of drugs by the formation of micelles owing to the action of its FA ingredients [5]. The quality of SO can affect safety and efficacy for clinical patients. There has been little research on the FA composition of SO, however, and there are quality control difficulties in the production of SO. It is important to establish qualitative and quantitative analytical methodology for determining the FA composition of SO.

To date, the methods used for separation and measurement of FAs are mainly chromatographic, including thin-layer chromatography [6], high-performance liquid chromatography [7,8], gas chromatography [9,10,11], supercritical fluid chromatography [12] and liquid chromatography with tandem mass spectrometry [13,14,15,16]. These methods cannot identify major chemical components rapidly and accurately. However, the gas chromatography/electron ionization-quadropole mass spectrometry (GC–EI-qMS) [17,18,19,20] technique coupled with the use of a professional database (NIST 08) can identify many compounds directly and accurately according to their fragment ions and abundance ratio [21,22,23]. In addition, GC–EI-qMS used in the selective ion monitoring (SIM) can identify target compounds rapidly and accurately despite interference from impurities [24], which is especially useful for the analysis of a lipid-based matrix, including SO.

FAs need to be derivatized before they can be analyzed by GC–MS because they have boiling points, which make gasification difficult. In many precolumn derivative methods [25], FAs are normally precolumn methylesterified (PME) into FA methyl esters (FAMEs) [26]. To ensure optimum conditions for methylesterification, the influence of important experimental parameters affecting the efficiency of methylesterification, including methyl reagent volume, temperature and time, were investigated using a Box–Behnken design (BBD) [27,28]. During optimization, the total peak area of the identified FAs was used to select the best conditions. This study developed and validated a method for the qualitative and quantitative profiling of the FA content in SOs for the first time.

SOs have been used widely in medicinal and culinary areas, but their authentication and standardization have encountered some problems owing to deliberate contamination with other animal or vegetable oils. It is difficult to identify the origins and species of SO accurately on the basis of appearance and morphology. Furthermore, SOs from different species or from different regions are not of uniform composition. In this study, a total of 18 batches of

SO collected from three provinces in China were analyzed by GC–EI-qMS to determine their FA compositions and principal component analysis (PCA) was used to evaluate and classify these samples.

2. RESULTS AND DISCUSSION

2.1. Optimal Results and Statistical Analysis of Precolumn Methylesterified (PME)

By retaining only the factors statistically significantly different at $p \leq 0.05$, the following final equation in terms of uncoded factors was obtained:

Total peak area = $+3.954 \times 10^9 + 9.987 \times 10^8 A + 1.196 \times 10^9 B + 9.163 \times 10^8 C + 1.099 \times 10^9 AB + 8.977 \times 10^8 AC + 9.306 \times 10^8 B - 1.596 \times 10^9 A^2 - 1.232 \times 10^9 B^2 - 1.452 \times 10^9 C^2$.

In all, 25 FA species can be identified from the chromatogram shown in Figure 1A. Comprehensive test results for response surface plots (3D) and contour plots (2D) show the total peak area was a maximum when the methylesterification conditions were: reagent volume 10 mL; temperature 60 °C; and time 10 min (Figure 1C).

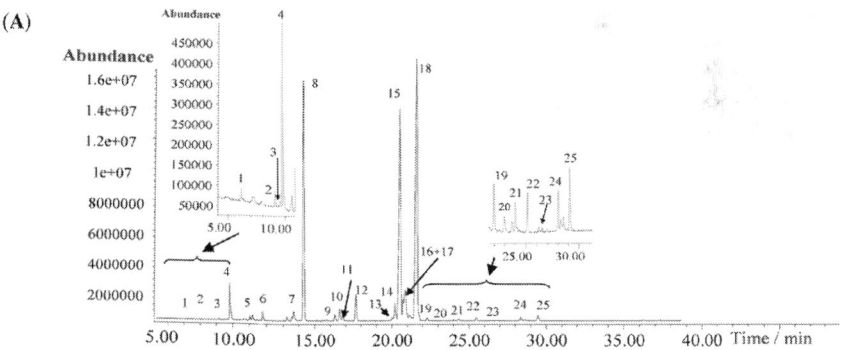

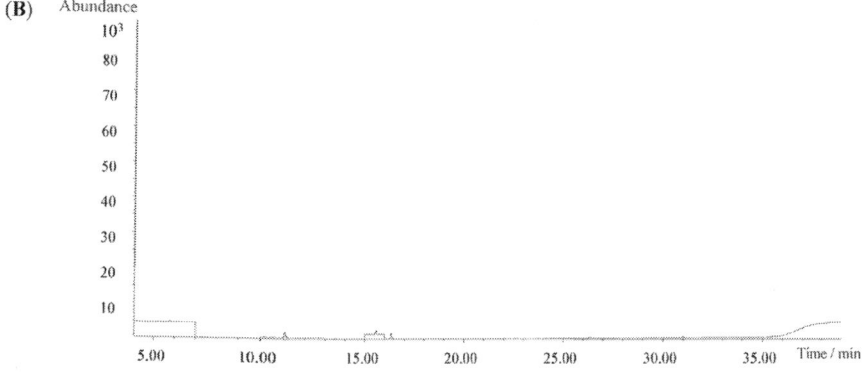

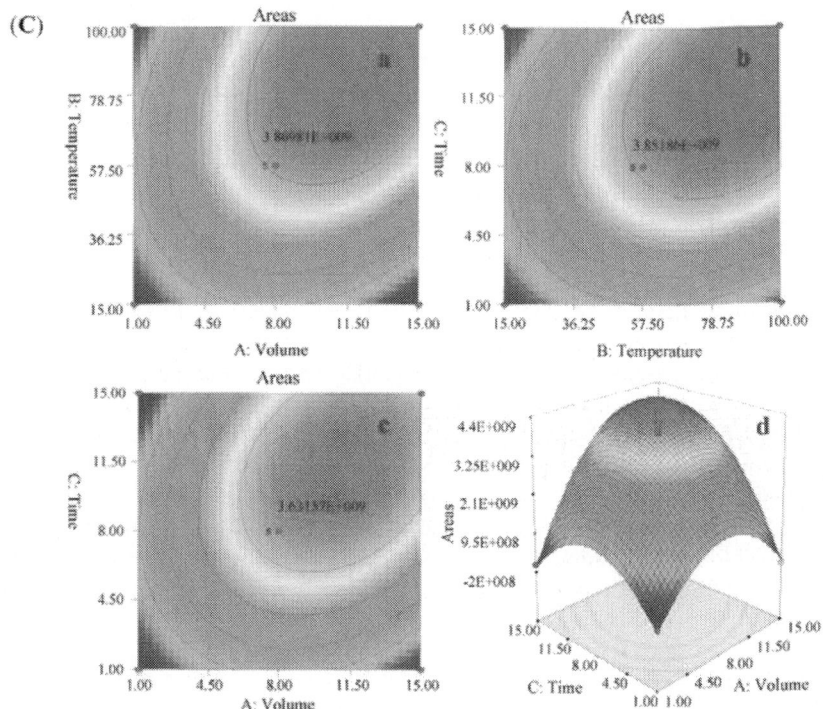

Figure 1. Optimization of precolumn methylesterified (PME) by Box–Behnken design (BBD)/GC–EI-qMS. (**A**) GC–EI-qMS chromatogram of the 25 fatty acid methyl esters (FAMEs) in suet oil (SO) sample under total ion chromatogram (TIC) mode. (**1**) Dodecanoic acid, methyl ester (DODME), (**2**) Methyl myristoleate, methyl ester, (**3**) Methyl 12-methyl-tridecanoate, methyl ester, (**4**) Tridecanoic acid, 12-methyl-, methyl ester, (**5**) Methyl tetradecanoate, methyl ester (MTEME), (**6**) Pentadecanoic acid, methyl ester (PENME), (**7**) (Z)-9-Hexadecenoic acid, methyl ester (9-HEME), (**8**) Hexadecanoic acid, methyl ester (HEXME), (**9**) Methyl 15-methylhexadecanoate, methyl ester, (**10**) cis-10-Heptadecenoic acid, methyl ester, (**11**) Heptadecanoic acid, methyl ester (HEPME), (**12**) (Z,Z)-9,12-Octadecadienoic acid, methyl ester (9,12-OCME), (**13**) Methyl 9-cis,11-trans-octadecadienoate methyl ester, (**14**) Methyl 10-trans,12-cis-octadecadienoate, (**15**) 9-Octadecenoic acid (E)-, methyl ester (9-OCME), (**16**) 9-Octadecenoic acid (Z)-, methyl ester, (**17**) 11-Octadecenoic acid, methyl ester, (**18**) Octadecanoic acid, methyl ester (OCTME), (**19**) cis-10-Nonadecenoic acid, methyl ester, (**20**) 10-Nonadecenoic acid, methyl ester, (**21**) Cyclopropaneoctanoic acid, 2-octyl-, methyl ester, (**22**) Nonadecanoic acid, methyl ester, (**23**) Methyl 8,11,14-eicosatrienoate, methyl ester, (**24**) cis-11-Eicosenoic acid, methyl ester, (**25**) Eicosanoic acid, methyl ester (EICME); (**B**) GC–EI-qMS chromatogram of representative blank samples under TIC mode; (**C**) Response surface plots (3-D) and contour (2-D) showing the total peaks areas with different methyl esterified condition. (**a**) 2-D panel of temperature-volum, (**b**) 2-D panel of time-temperature, (**c**) 2-D panel of time-volum, (**d**) 3-D response surface plots.

2. RESULTS AND DISCUSSION

2.2. Fatty Acids (FAs) Composition in Suet Oil (SO)

Identification of FAs was achieved by comparing molecular mass, ion fragments and abundance ratios in the NIST 08 spectral library. A typical total ion chromatogram obtained for SO samples is shown in Figure 1A.

The FAs in SO were investigated using optimized PME: 10 mL of BF_3–MeOH (14%, v/v), 60 °C and 10 min. The PME/GC–EI-qMS analysis of SO led to the identification of 25 different FAs (Table 1): including saturated FAs (dodecanoic acid, 12-methyl-tridecanoate, tridecanoic acid, 12-methyl-tetradecanoate, pentadecanoic acid, hexadecanoic acid, heptadecanoic acid, octadecanoic acid, cyclopropaneoctanoic acid, nonadecanoic acid, eicosanoic acid, cis-11-eicosenoic acid and 8,11,14-eicosatrienoate) and unsaturated FAs (myristoleate, 9-hexadecanoic acid, cis-10-heptadecanoic acid, (Z,Z)-9,12-octadecadienoic acid, 10-nonadecanoic acid, 10-trans,12-cis-octadecadienoate, (E)-9-octadecanoic acid, (Z)-9-octadecanoic acid, 11-octadecanoic acid, cis-10-nonadecanoic acid and 9-cis,11-trans-octadecadienoate).

In all, 25 FAs were identified (match > 90%). Hexadecanoic acid, octadecanoic acid and (E)-9-octadecanoic acid were the three most abundant and occupied 16.46%, 37.96% and 19.47% of the total peak area, respectively (Table 1).

2.3. Validation of Quantitative Analysis

Methylester derivatives of the ten most abundant FAs were purchased for use as standards. These FAs in the SO samples were quantified by derivatization into FAMEs, which were analyzed by GC–EI-qMS without significant matrix interference. Four fragment ions were monitored in the SIM mode for each compound. The best characteristic ion in the spectrum was selected for quantification of each FAME and the other three were used for confirmation.

The validity of the method was investigated by examination of the linearity, recovery, and limit of quantification (LOQ) for all FAMEs in this study. The ranges of concentration, regression equations, r^2 (coefficient of determination), recovery, relative standard deviation (RSD) and LOQ for the target FAMEs are given in Table 2. Most of the FAMEs had good linearity ($r^2 > 0.999$); more importantly, the results showed a stabilized recovery of ten FAMEs in the range 82.1%–98.7% with optimized PME parameters (Table 2, Figure S1) and the LOQ values of these FAMEs ranged from 1.25 to 5.95 µg/L. These results indicated LOQ was sufficiently low to meet the requirements of determination of the FA composition of SOs. A chromatogram of these ten FAMEs in SOs obtained in the SIM mode is shown in Figure 2A.

Table 1. Fatty acids identified in SO sample using precolumn esterified/GC–EI-qMS.

No.	RT (min)	Compounds Name	CAS No.	Mw[a]	Formula	Match (%)	RC[b] (%)
1	6.142	Dodecanoic acid, methyl ester	000111-82-0	214	$C_{13}H_{26}O_2$	98	0.11
2	9.016	Methyl myristoleate	056219-06-8	240	$C_{15}H_{28}O_2$	96	0.19
3	9.504	Methyl 12-methyl-tridecanoate	1000336-46-9	242	$C_{15}H_{30}O_2$	98	0.087
4	9.739	Tridecanoic acid, 12-methyl-, methyl ester	005129-58-8	242	$C_{15}H_{30}O_2$	94	0.25
5	11.071	Methyl tetradecanoate	000124-10-7	242	$C_{15}H_{30}O_2$	98	2.52
6	11.663	Pentadecanoic acid, methyl ester	007132-64-1	256	$C_{16}H_{32}O_2$	99	0.29
7	13.614	9-Hexadecenoic acid, methyl ester, (Z)-	001120-25-8	268	$C_{17}H_{32}O_2$	99	2.51
8	14.311	Hexadecanoic acid, methyl ester	000112-39-0	270	$C_{17}H_{34}O_2$	98	16.46
9	16.253	Methyl 15-methylhexadecanoate	1000336-34-2	284	$C_{18}H_{36}O_2$	99	0.64
10	16.531	cis-10-Heptadecenoic acid, methyl ester	1000333-62-1	282	$C_{18}H_{34}O_2$	99	0.98
11	16.723	Heptadecanoic acid, methyl ester	001731-92-6	284	$C_{18}H_{36}O_2$	99	1.06
12	17.507	9,12-Octadecadienoic acid (Z,Z)-, methyl ester	000112-63-0	294	$C_{19}H_{34}O_2$	99	1.61
13	19.492	Methyl 9-cis,11-trans-octadecadienoate	1000336-44-0	294	$C_{19}H_{34}O_2$	95	1.29
14	19.919	Methyl 10-trans,12-cis-octadecadienoate	1000336-44-2	294	$C_{19}H_{34}O_2$	96	0.10
15	20.119	9-Octadecenoic acid (E)-, methyl ester	001937-62-8	296	$C_{19}H_{36}O_2$	99	37.96
16	20.546	9-Octadecenoic acid (Z)-, methyl ester	000112-62-9	296	$C_{19}H_{36}O_2$	99	1.97
17	20.955	11-Octadecenoic acid, methyl ester	052380-33-3	296	$C_{19}H_{36}O_2$	99	2.05
18	21.992	Octadecanoic acid, methyl ester	000112-61-8	298	$C_{19}H_{38}O_2$	99	19.47
19	22.166	cis-10-Nonadecenoic acid, methyl ester	1000333-64-4	310	$C_{20}H_{38}O_2$	98	0.065
20	23.316	10-Nonadecenoic acid, methyl ester	056599-83-8	310	$C_{20}H_{38}O_2$	93	0.24
21	24.308	Cyclopropaneoctanoic acid, 2-octyl-, methyl ester	3971-54-8	310	$C_{20}H_{38}O_2$	99	0.42
22	25.287	Nonadecanoic acid, methyl ester	001731-94-8	312	$C_{20}H_{40}O_2$	99	0.26
23	26.829	Methyl 8,11,14-eicosatrienoate	1000336-38-1	320	$C_{21}H_{36}O_2$	97	0.034
24	28.262	cis-11-Eicosenoic acid, methyl ester	1000333-63-8	324	$C_{21}H_{40}O_2$	99	0.12
25	29.653	Eicosanoic acid, methyl ester	001120-28-1	326	$C_{21}H_{42}O_2$	99	0.20

[a] Mw = Molecular Weight (nominal values); [b] RC (%) = The relative content of total peak areas, the sum of the RC was 90.89%, the other 9.11% may contain inorganic elements, glycerin and something else.

Table 2. The linear regression equations, the correlation coefficient (r), limit of quantification (LOQ), recoveries of 10 fatty acids under GC–EI-qMS selective ion monitoring (SIM) conditions.

NO. of Identified Fatty Acids	Compounds	Linear Regression Equations	Coefficient of Determination/r^2	Linear Range μg/ml	Qualitative/ Quantitative Ions	Abundance Ratio (%)	0.5 Times Spiked ($n=3$)		1.0 Times Spiked ($n=3$)		2.0 Times Spiked ($n=3$)		LOQ (μg/ml, ×10[-2])
							Recovery (%)	RSDs (%)	Recovery (%)	RSDs (%)	Recovery (%)	RSDs (%)	
1	DODME	$Y = 1.56 \times 10^4 X - 1.79 \times 10^2$	0.999	0.010–10.0	74*:28:87:214	100:8:64:6	85.3	4.3	92.4	4.8	93.2	2.2	1.25
5	MTEME	$Y = 1.54 \times 10^4 X - 1.48 \times 10^3$	0.999	0.013–12.8	74*:87:143:199	100:58:24:16	95.2	4.1	97.2	3.3	98.7	4.5	1.60
6	PENME	$Y = 1.86 \times 10^4 X - 4.74 \times 10^3$	0.999	0.011–12.0	74*:87:143:213	100:68:20:18	91.3	8.2	92.8	4.5	95.8	4.2	1.40
7	9-HEXME	$Y = 3.10 \times 10^5 X - 1.13 \times 10^5$	0.999	0.020–20.0	55*:74:87:236	100:66:50:23	83.6	4.8	94.6	3.5	96.6	2.4	2.50
8	HEXME	$Y = 2.62 \times 10^4 X - 1.579 \times 10^4$	0.999	0.048–50.0	74*:87:143:227	100:70:20:14	87.5	5.2	95.4	5.4	96.9	3.8	5.95
11	HEPME	$Y = 2.93 \times 10^4 X - 6.54 \times 10^3$	0.999	0.013–13.0	74*:87:143:241	100:70:22:15	87.6	6.8	90.8	5.0	92.7	4.1	1.63
12	9,12-OCME	$Y = 7.71 \times 10^3 X - 1.15 \times 10^2$	0.999	0.020–20.0	67*:81:95:294	100:92:66:16	91.7	4.4	91.7	2.8	93.9	3.6	2.50
15	9-OCME	$Y = 5.02 \times 10^3 X - 3.62 \times 10^2$	0.999	0.020–20.0	55*:41:81:222	100:62:40:24	84.8	5.1	88.4	4.2	97.2	4.4	2.50
18	OCTME	$Y = 3.31 \times 10^4 X - 1.24 \times 10^4$	0.999	0.030–32.0	74*:87:143:255	100:72:23:14	85.5	4.3	86.7	5.2	88.8	5.1	3.75
25	EICME	$Y = 3.18 \times 10^4 X - 4.78 \times 10^3$	0.999	0.010–10.8	74*:87:143:255	100:76:26:18	82.1	3.9	90.1	4.7	89.4	3.7	1.25

* Quantitative ion, LOQ was calculated as 10 times of the signal to noise ratio (10 S/N). 0.5, 1.0 and 2.0 times "Spiked" means the added standards amount was 0.5, 1.0 and 2.0 times of the initial content of samples.

2. RESULTS AND DISCUSSION 249

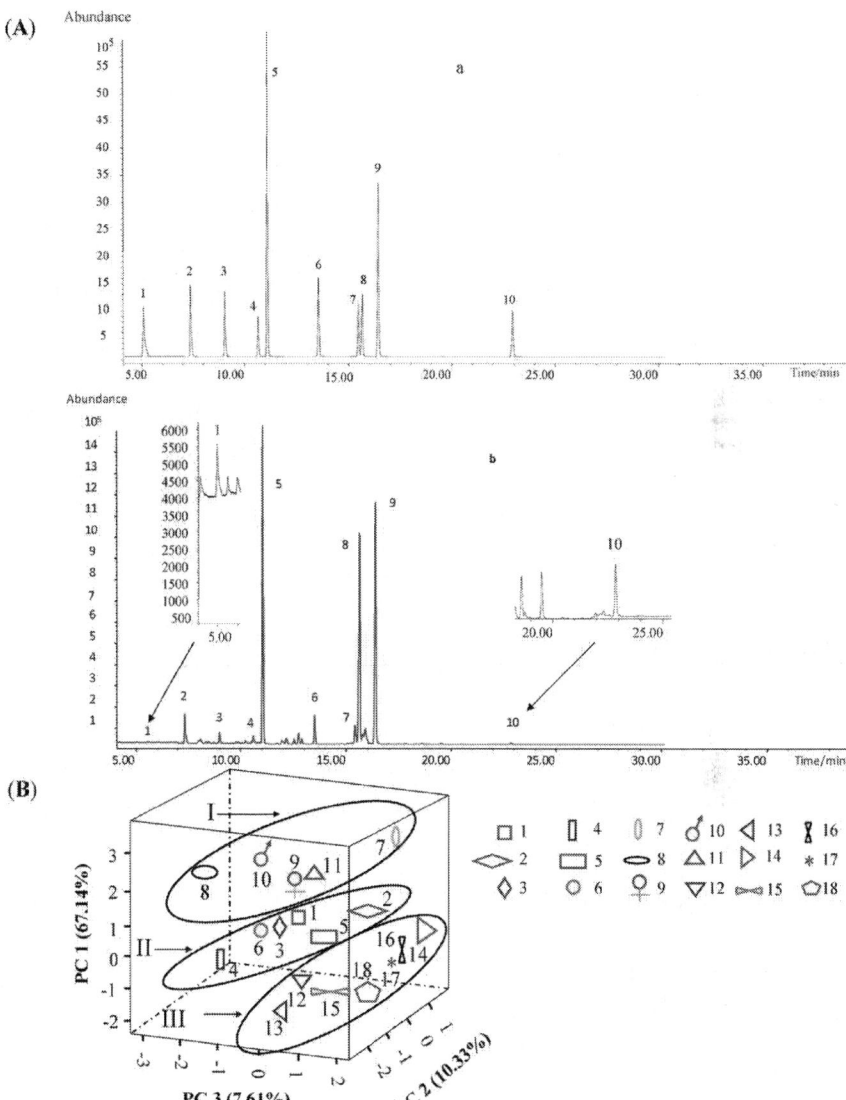

Figure 2. GC–EI-qMS chromatograms of the ten FAMEs standard mixture and sample under SIM mode and principal component analysis (PCA) of 18 SO samples. (**A**) (**a**) mixed standard solution (the concentration of (**1–10**) was 10.8, 47.6, 31.6, 13.0, 11.2, 12.8, 10.0, 20.0, 20.0 and 20.0 μg/mL, respectively), (**b**) SO sample solution. (**1**) DODME, (**2**) MTEME, (**3**) PENME, (**4**) 9-HEXME, (**5**) HEXME, (**6**) HEPME, (**7**) 9,12-OCME, (**8**) 9-OCME, (**9**) OCTME, (**10**) EICME; (**B**) The 3D scatter plot obtained by PCA of 18 SO samples.

2.4. Quantitative Results

The molecular species detected after methylesterification were FAMEs rather than FAs. Therefore, in order to quantify the FAs, the contents of FAMEs were converted into FAs by multiplication with the corresponding coefficient as follows:

$$W_{\text{FAs}} (\%) = A \times W_{\text{FAMEs}} (\%) \tag{1}$$

where W is concentration and A is the molecular mass of the FAMEs/the molecular mass of the FAs.

The results for ten FAs in 18 batches of SO are given in Table 3. The contents of hexadecanoic acid and octadecanoic acid were 3.79%–13.22% and 3.41%–18.11%, respectively. The contents of (E)-9-octadecanoic acid in SO batches 3–6, 8, 12–15 and 18 were 4.48%–6.28%, second only to hexadecanoic acid and octadecanoic acid, whereas no other FA was detected. On the basis of the quantitative results for 18 batches of SO, the content of octadecanoic acid was greatest, followed by hexadecanoic acid, (E)-9-octadecanoic acid and tetradecanoate in that order.

2.5. Principal Component Analysis (PCA) of the SO Samples

In order to evaluate the variation between batches of SO, PCA was applied on the basis of the contents of the ten most abundant FAs. The first three principal components (PC1, PC2 and PC3) with >85% of the whole variance were extracted for analysis. PC1, PC2 and PC3 accounted for 67.14%, 10.33% and 7.61% of the total variance, respectively (Table S1). The remaining principal components had only a minor effect on the model and were discarded. The component loading matrix is given in Table S2 and Table S3. According to the loadings, PC1 had a good correlation with each of the ten FA compounds. The results mentioned above suggested that most of the compounds contributed to the classification of the samples. The scatter plots are shown in Figure 2B, where each sample is represented as one marker.

The dots of 18 samples were classified into group I, group II or group III in accord with their origin. Dots in groups II and III were relatively close to each other, indicating a close relationship among the six batches from Anhui and the seven batches from Jiangsu. The dots in group I were quite scattered, suggesting diversification of the five batches from Qinghai Province. These observations might be explained as follows. Firstly, the land area of Qinghai Province (722,300 km^2) is larger compared to Anhui Province (1,396,002 km^2) and Jiangsu Province (106,700 km^2), representing a greater area for diversity of the samples. Secondly, Qinghai, Anhui and Jiangsu provinces are considerably different environments with large differences in climate, which influences the differences in FA metabolism in domestic sheep and goats. Thirdly, Anhui and Jiangsu provinces are geographic neighbors, which is reflected in the similarities among samples from these two origins. Finally, the SO samples from Qinghai Province were from sheep, whereas those obtained from Anhui and Jiangsu provinces were from goats.

Table 3. Contents of ten FAs in 18 batches of Suet oil ($n = 3$).

FAs Compounds	DODME	MTEME	PENME	9-HEXME	HEXME	HEPME	9,12-OCME	9-OCME	OCTME	EICME
Coefficient "A" [a]	0.9348	0.9414	0.9454	0.9478	0.9482	0.9508	0.9524	0.9527	0.9531	0.9571
Batch					Content (g/100 g, %) [b]					
1	0.0240 ± 0.0014	1.1476 ± 0.0014	0.2124 ± 0.0002	0.5716 ± 0.0012	8.9067 ± 0.1483	0.4996 ± 0.0006	0.8667 ± 0.0015	Nd	12.5609 ± 0.1614	0.0933 ± 0.0021
2	0.0284 ± 0.0010	1.0827 ± 0.0012	0.1787 ± 0.0015	0.4596 ± 0.0029	7.6098 ± 0.0040	0.5022 ± 0.0011	1.1867 ± 0.0023	Nd	11.0827 ± 0.0011	0.1040 ± 0.0023
3	0.0284 ± 0.0009	0.8684 ± 0.0010	0.2009 ± 0.0028	0.4044 ± 0.0025	6.2578 ± 0.0157	0.5289 ± 0.0053	1.1218 ± 0.0102	Nd	8.7600 ± 0.3606	0.0942 ± 0.0013
4	0.0196 ± 0.0004	0.4862 ± 0.0015	0.2098 ± 0.0039	0.4204 ± 0.0023	4.4382 ± 0.0011	0.4551 ± 0.0022	0.6587 ± 0.0065	Nd	4.0124 ± 0.0017	0.0471 ± 0.0008
5	0.0231 ± 0.0023	0.7707 ± 0.0083	0.1991 ± 0.0017	0.3191 ± 0.0017	5.8071 ± 0.0045	0.5707 ± 0.0015	0.6604 ± 0.0024	4.5991 ± 0.0053	10.2213 ± 0.0163	0.1467 ± 0.0032
6	0.0418 ± 0.0012	1.1912 ± 0.0034	0.2844 ± 0.0012	1.3378 ± 0.0051	7.5067 ± 0.0074	0.6187 ± 0.0052	0.9227 ± 0.0008	Nd	8.3093 ± 0.0297	0.0978 ± 0.0017
7	0.0400 ± 0.0031	1.9422 ± 0.0068	0.3040 ± 0.0108	1.1564 ± 0.0020	13.2151 ± 0.0241	0.7956 ± 0.0023	1.4640 ± 0.0028	Nd	18.1129 ± 0.0028	0.1662 ± 0.0035
8	0.0328 ± 0.0017	1.1182 ± 0.0037	0.2729 ± 0.0017	0.7662 ± 0.0026	7.9458 ± 0.0026	0.7218 ± 0.0017	0.9048 ± 0.0033	Nd	12.1111 ± 0.0015	0.1582 ± 0.0001
9	0.0400 ± 0.0031	1.4516 ± 0.0012	0.3902 ± 0.0035	0.8649 ± 0.0033	10.8133 ± 0.0034	1.1004 ± 0.0001	1.4924 ± 0.0035	Nd	12.9360 ± 0.0275	0.1111 ± 0.0016
10	0.0280 ± 0.0046	1.0978 ± 0.0045	0.2516 ± 0.0024	0.4222 ± 0.0029	8.264 ± 0.0394	0.6960 ± 0.0027	1.0480 ± 0.0042	Nd	13.0320 ± 0.0337	0.1467 ± 0.0040
11	0.0373 ± 0.0035	1.1991 ± 0.0031	0.4791 ± 0.0033	0.4764 ± 0.0040	10.0676 ± 0.0045	0.9218 ± 0.0017	0.8107 ± 0.0059	Nd	14.2569 ± 0.0029	0.1040 ± 0.0016
12	0.0356 ± 0.0039	1.0196 ± 0.0041	0.2658 ± 0.0039	0.7564 ± 0.0031	6.7449 ± 0.0017	0.4942 ± 0.0033	0.7787 ± 0.0041	6.2764 ± 0.0023	6.0773 ± 0.0036	0.0560 ± 0.0034
13	0.0178 ± 0.0040	0.4080 ± 0.0046	0.1769 ± 0.0034	0.3396 ± 0.0044	3.7902 ± 0.0047	0.3804 ± 0.0018	0.5449 ± 0.0039	5.4489 ± 0.0048	3.4080 ± 0.0051	0.0382 ± 0.0012
14	0.0322 ± 0.0035	1.1831 ± 0.0040	0.2391 ± 0.0045	0.4276 ± 0.0046	8.3760 ± 0.0164	0.5822 ± 0.0029	1.2240 ± 0.0167	5.2880 ± 0.0282	11.2276 ± 0.0060	0.087 ± 0.0045
15	0.0240 ± 0.0023	0.6462 ± 0.0034	0.1458 ± 0.0028	0.4587 ± 0.0013	4.6827 ± 0.0034	0.3671 ± 0.0035	0.6240 ± 0.0293	5.0649 ± 0.0034	6.6560 ± 0.0220	0.0773 ± 0.0038
16	0.0267 ± 0.0029	0.8071 ± 0.0039	0.1636 ± 0.0032	0.3363 ± 0.0032	5.8738 ± 0.0015	0.4747 ± 0.0038	1.0738 ± 0.0028	4.8996 ± 0.0044	9.0898 ± 0.0042	0.0969 ± 0.0023
17	0.0267 ± 0.0034	0.7911 ± 0.0061	0.1564 ± 0.0028	0.3209 ± 0.0078	5.4720 ± 0.0406	0.4329 ± 0.0030	1.0169 ± 0.0033	4.6578 ± 0.0046	8.1529 ± 0.0040	0.0862 ± 0.0039
18	0.0160 ± 0.0051	0.6969 ± 0.0034	0.0907 ± 0.0064	0.3413 ± 0.0031	5.5209 ± 0.0071	0.2480 ± 0.0046	0.5013 ± 0.0034	4.4756 ± 0.0034	7.9644 ± 0.0033	0.0622 ± 0.0042

[a] A = Molecular Weight $_{(FAMEs)}$/Molecular Weight $_{(FAs)}$; [b] W_{FAs} (%) = A × W_{FAMEs} (%); Nd = not detected.

3. EXPERIMENTAL SECTION

3.1. Materials

Methyl dodecanoic acid (DODME ≥ 98.0 (purity)), methyl tetradecanoate (MTEME ≥ 99.0), methyl pentadecanoic acid (PENME ≥ 98.0), methyl 9-hexadecenoic acid (Z) (9-HEME ≥ 99.0), methyl hexadecanoic acid (HEXME

≥ 99.0), methyl heptadecanoic acid (HEPME ≥ 99.0), 9,12-methyl octadecadienoic acid (Z,Z) (9,12-OCME ≥ 99.0), methyl 9-octadecenoic acid (E) (9-OCME ≥ 99.0), methyl octadecanoic acid (OCTME ≥ 98.0), methyl eicosanoic acid (EICME ≥ 99.0), boron trifluoride-methanol (14%, v/v), sodium hydroxide (NaOH) and sodium (NaCl) were purchased from Anpel Scientific Instrument Co., Ltd. (Shanghai, China). HPLC grade methanol, and n-hexane were obtained from Merck (Darmstadt, Germany).

3.2. Sample Material

Eighteen batches of SO samples were purchased from Qinghai (batches 7–11), Jiangsu (batches 12–18) and Anhui (batches 1–6) provinces, China between January and July 2013 (Figure S2). All samples were stored in darkness at temperatures <4 °C. For the blank sample, the n-hexane was used instead of SOs.

3.3. Box–Behnken Design for Optimization of PME Parameters

The application of an effective PME methodology requires optimization of the main parameters that influence the methylesterification process, including the volume of methyl reagent, temperature and time.

A Box–Behnken Design, a response surface methodology, was used in this study. Design Expert 7.0.0 software was used for analyzing the experimental data. The study type was Response Surface, the initial design was Box–Behnken, the design model was Quadratic and Blocks was No Blocks. A Box–Behnken statistical screening design with three independent variables (A, PME volume; B, PME temperature; C, PME time) was used to optimize the PME process for the qualitative and quantitative analysis. Statistically significant difference was set at $p \leq 0.05$. The r^2 value of the "Final Equation" > 0.995 indicated derived results were accurate. Data were expressed as mean ± standard deviation (SD) of triplicate determinations. Statistical calculations used Statistical Product and Service Solutions (SPSS) version 16.0 software (SPSS Inc., Chicago, IL, USA). One-way ANOVA was used for evaluating the statistical differences among samples.

3.4. PME Procedure

A 0.4 g sample of SO from batch 12 was weighed and placed into a 50-mL conical flask followed by 15 mL NaOH–MeOH (0.5 mol/L) then heated at 60 °C in a waterbath for 20 min until the yellow beads of SO disappeared completely after cooling. The flask contents were subjected to the PME procedure, in which 10 mL of boron trifluoride methanol (BF_3–MEOH, 14%v/v) was added to the flask then heated at 60 °C in a waterbath for 10 min. The mixture was cooled and then 10 mL of n-hexane and 10 mL of saturated NaCl were added. Samples 1.5 mL of supernatants were injected through a 0.45-μm pore size membrane before GC–EI-qMS qualitative analysis.

3.5. Sample Pretreatment for Quantitative Analysis
Eighteen batches of SO were treated as described in section 3.4 above. Sequentially, 25-μL was transferred into a 10-mL volumetric flask followed by addition of n-hexane to a final volume of 10 mL and then shaken. After passage through an organic 0.45-μm pore size filter, the treated samples were injected into the GC–EI-qMS for quantitative analysis.

3.6. Preparation of Standard Solutions
Stock solutions of the ten FAMEs (DODME, MTEME, PENME, 9-HEXME, HEXME, HEPME, 9,12-OCME, EICME, OCTME and 9-OCME) were prepared in n-hexane at concentrations of 20.0, 11.2, 20.0, 12.8, 31.6, 47.6, 10.8, 20.0, 13.0 and 10.0 μg/mL, respectively. Appropriate amounts of the above stock solutions were mixed and diluted into a series of concentrations with n-hexane to obtain the working solutions. All solutions were stored at <4 °C.

3.7. GC–EI-qMS Analysis Conditions
For separation, detection and identification of FAs, the qualitative and quantitative analyses were made with a GC–EI-qMS instrument (Agilent 7890/5975) coupled to an automatic sampler (Agilent 7693) and an electron impact ionization source (Agilent, Santa Clara, CA, USA). Water was purified by a Milli-Q Plus apparatus (Millipore, Bedford, MA, USA). The H2050R centrifugal apparatus was provided by the Hunan Saite xiangyi centrifuge instrument Co., Ltd. (Xiangya, China).

Analytes were separated using a 30 m × 0.25 mm capillary column (HP-5 ms 0.25 μm film thickness; Agilent Technology, Santa Clara, CA, USA). The primary oven temperature protocol was: 150 °C for 1 min; increased to 200 °C at 5 °C/min; maintained at this temperature for 5 min; increased to 250 °C at a rate of 5 °C /min; maintained at this temperature for 5 min; increased to 300 °C at a rate of 5 °C/min; and maintained at this temperature for 10 min. The injection port temperature was 250 °C. The carrier gas was helium at a constant flow of 1 mL/min. The MS operating conditions in the splitless injection mode were as follows: ion source temperature 280 °C; electron energy 70 eV; emission current 250 μA; injection volume 0.2 μL; and solvent delay 4 min. The SIM mode was used for quantitative determination of FAs.

3.8. Method for PCA of Samples
PCA was done with SPSS 16.0 software (SPSS, Chicago, IL, USA) [29]. In this study, the contents of the ten FAs in the 18 SO samples were used as a data matrix with 18 rows and ten columns for PCA analysis after normalization. The first three PCs were extracted, and the scatter plot was obtained by plotting the scores of PC1 $vs.$ PC2 and PC3.

4. CONCLUSIONS

The optimal conditions for methylesterification of FAs were obtained by a Box–Behnken Design, which identified 25 kinds of FAs in SO by GC–EI-qMS. In addition, ten FAs in 18 batches of SO were analyzed with good performance with regard to selectivity, recovery, precision and accuracy. Significant differences among origins in FA composition profiles and their contents were revealed. The method described here could be used in quality control and standardization of SOs and their products as well as providing supportive chemical information.

ACKNOWLEDGMENTS

This work was supported by the Natural Science Foundation of China (No. 81274088).

AUTHOR CONTRIBUTIONS

Jun Jiang designed research; Jun Jiang and Xiaobin Jia performed research and analyzed the data; Jun Jiang wrote the paper. Both authors read and approved the final manuscript.

REFERENCES

1. Thurnhofer, S.; Hottinger, G.; Vetter, W. Enantioselective determination of anteiso fatty acids in food samples. *Anal. Chem.* **2007**, *79*, 4696–4701.
2. Mattacks, C.A.; Sadler, D.; Pond, C.M. Site-specific differences in the action of NRTI drugs on adipose tissue incubated *in vitro* with lymphoid cells, and their interaction with dietary lipids. *Comp. Biochem. Phys.* **2003**, *135*, 11–29.
3. Thurnhofer, S.; Vetter, W. A gas chromatography/electron ionization-mass spectrometry-selected ion monitoring method for determining the fatty acid pattern in food after formation of fatty acid methyl esters. *J. Agric. Food Chem.* **2005**, *53*, 8896–8903.
4. Barzanti, V.; Maranesi, M.; Cornia, G.L.; Malavolti, M.; Mordenti, T.; Pregnolato, P. Effect of dietary oils containing different amounts of precursor and derivative fatty acids on prostaglandin E2 synthesis in liver, kidney and lung of rats. *Prostag. Leukotr. Essent.* **1999**, *60*, 49–54.

REFERENCES

5. Cui, L.; Sun, E.; Zhang, Z.H.; Tan, X.B.; Wei, Y.J.; Jin, X. Enhancement of epimedium fried with suet oil based on *in vivo* formation of self-assembled flavonoid compound nanomicelles. *Molecules* **2012**, *17*, 12984–12996.
6. Ansorena, D.; Raes, K.; de Smet, S.; Demeyer, D. Analysis of fatty acid isomers in ruminant tissues by silver thin layer chromatography followed by gas chromatography. *Meded. Rijksuniv. Gent Fak. Landbouwkd. Toegep. Biol. Wet.* **2001**, *66*, 365–372.
7. Zhang, S.; Sun, Y.; Sun, Z.; Wang, X.; You, J.; Suo, Y. Determination of triterpenic acids in fruits by a novel high performance liquid chromatography method with high sensitivity and specificity. *Food Chem.* **2014**, *146*, 264–269.
8. Wang, A.; Li, G.; You, J.; Ji, Z. A new fluorescent derivatization reagent and its application to free fatty acid analysis in pomegranate samples using HPLC with fluorescence detection. *J. Sep. Sci.* **2013**, *36*, 3853–3859.
9. Bielawska, K.; Dziakowska, I.; Roszkowska-Jakimiec, W. Chromatographic determination of fatty acids in biological material. *Toxicol. Mech. Methods* **2010**, *20*, 526–537.
10. Li, A.; Ha, Y.; Wang, F.; Li, W.; Li, Q. Determination of thermally induced trans-fatty acids in soybean oil by attenuated total reflectance fourier transform infrared spectroscopy and gas chromatography analysis. *J. Agric. Food Chem.* **2012**, *60*, 10709–10713.
11. Bogusz, S.J.; Hantao, L.W.; Braga, S.C.; de Matos França, V.C.; da Costa, M.F. Solid-phase microextraction combined with comprehensive two dimensional gas chromatography for fatty acid profiling of cell wall phospholipids. *J. Sep. Sci.* **2012**, *35*, 2438–2444.
12. Hori, K.; Matsubara, A.; Uchikata, T.; Tsumura, K.; Fukusaki, E.; Bamba, T. High-throughput and sensitive analysis of 3-monochloropropane-1,2-diol fatty acid esters in edible oils by supercritical fluid chromatography/tandem mass spectrometry. *J. Chromatogr. A* **2012**, *1250*, 99–104.
13. Aslan, M.; Ozcan, F.; Aslan, B.; Yücel, G. LC–MS/MS analysis of plasma polyunsaturated fatty acids in type 2 diabetic patients after insulin analog initiation therapy. *Lipidis Health Dis.* **2013**, *12*, 169.
14. Derogis, P.B.; Freitas, F.P.; Marques, A.S.; Cunha, D.; Appolinário, P.P.; de Paula, F. Detection and quantification of Hydroperoxy and Hydroxydo-cosahexaenoic acids as a tool for lipidomic analysis. *PLoS One* **2013**, *8*, e77561.
15. Le Faouder, P.; Baillif, V.; Spreadbury, I.; Motta, J.P.; Rousset, P.; Chêne, G. LC–MS/MS method for rapid and concomitant quantification

of pro-inflammatory and pro-resolving polyunsaturated fatty acid metabolites. *J. Chromatogr. B* **2013**, *932*, 123–133.
16. Takahashi, H.; Suzuki, H.; Suda, K.; Yamazaki, Y.; Takino, A.; Kim, Y.I. Long-chain free fatty acid profiling analysis by liquid chromatography–mass spectrometry in mouse treated with peroxisome proliferator-activated receptor α agonist. *Biosci. Biotechnol. Biochem.* **2013**, *77*, 2288–2293.
17. Zeng, A.X.; Chin, S.T.; Nolvachai, Y.; Kulsing, C.; Sidisky, L.M.; Marriott, P.J. Characterisation of capillary ionic liquid columns for gas chromatography mass spectrometry analysis of fatty acid methyl esters. *Anal. Chim. Acta* **2013**, *803*, 166–173.
18. Valianpour, F.; Selhorst, J.J.; van Lint, L.E.; van Gennip, A.H.; Wanders, R.J.; Kemp, S. Analysis of very long-chain fatty acids using electrospray ionization mass spectrometry. *Mol. Genet. Metab.* **2003**, *79*, 189–196.
19. Byss, M.; Tríska, J.; Elhottová, D. GC–MS–MS analysis of bacterial fatty acids in heavily creosote-contaminated soil samples. *Anal. Bioanal. Chem.* **2007**, *387*, 1573–1577.
20. Oursel, D.; Loutelier-Bourhis, C.; Orange, N.; Chevalier, S.; Norris, V.; Lange, C.M. Identification and relative quantification of fatty acids in *Escherichia coli* membranes by gas chromatography/mass spectrometry. *Rapid Commun. Mass Spectrom.* **2007**, *21*, 3229–3233.
21. Catarina, L.S.; José, S.C. Profiling of volatiles in the leaves of Lamiaceae species based on headspace solid phase microextraction and mass spectrometry. *Food Res. Int.* **2013**, *51*, 378–387.
22. Mahinda, W.; Thava, V.; Feral, T.; Kevin, S. Volatile flavour composition of cooked by-product blends of chicken, beef and pork: A quantitative GC–MS investigation. *Food Res. Int.* **2001**, *34*, 149–158.
23. Diana, A.; Olga, G.; Iciar, A.; José, B. Analysis of volatile compounds by GC–MS of a dry fermented sausage: Chorizo de Pamplona. *Food Res. Int.* **2001**, *34*, 67–75. Dodds, E.D.; McCoy, M.R.; Rea, L.D.; Kennish, J.M. Gas chromatographic quantification of fatty acid methyl esters: Flame ionization detection *vs.* electron impact mass spectrometry. *Lipids* **2005**, *40*, 419–428.
24. Saliu, F.; Orlandi, M. *In situ* alcoholysis of triacylglycerols by application of switchable-polarity solvents. A new derivatization procedure for the gas chromatographic analysis of vegetable oils. *Anal. Bioanal. Chem.* **2013**, *405*, 8677–8684.
25. Igarashi, M.; Tsuzuki, T.; Kambe, T.; Miyazawa, T. Recommended methods of fatty acid methylester preparation for conjugated dienes and trienes in food and biological samples. *J. Nutr. Sci. Vitaminol.* **2004**, *50*, 121–128.

26. Box, G.E.P.; Wlson, K.B. On the experimental attainment of optimum conditions. *J. R. Stat. Soc.* **1951**, *13*, 1–45.
27. Luo, C.; Chen, Y.S. Optimization of extraction technology of Se-enriched *Hericium erinaceum* polysaccharides by Box–Behnken statistical design and its inhibition against metal elements loss in skull. *Carbohydr. Polym.* **2010**, *82*, 845–860.
28. Jiang, J.; Feng, L.; Li, J.; Sun, E.; Ding, S.M.; Jia, X.B. Multielemental composition of suet oil based on quantification by ultrawave/ICP-MS coupled with chemometric analysis. *Molecules* **2014**, *19*, 4452–4465.

Index

3-Hydroxypyridine-4-one, 43

A
Amino acids, 27, 35
Antioxidant prop1erties, 27
ATR-FTIR, 15, 16, 17, 18, 19, 22, 24, 215, 216, 217, 218, 220, 221, 222, 224, 226, 228

B
Barley, 63, 64, 77
Bone Tissue, 99, 111

C
Cellular Transformations, 158
Chemometric Method, 231
Chemometrics, 1, 13, 15, 25, 26, 27, 43, 45, 60, 61, 63, 81, 96, 97, 99, 110, 111, 113, 121, 122, 123, 141, 142, 143, 144, 145, 146, 149, 199, 202, 227, 228, 242, 243
Composition profiles, 243

D
Data Mining, 15, 228

E
Echinacea, 179, 180, 181, 182, 184, 185, 186, 187, 188, 189, 190, 191, 192, 193, 194, 195, 197, 198, 199

F
Fatty acids, 243, 248
Feature Selection, 60, 135, 215
Forensic Science, 99, 108, 109, 110
Fourier Transform Infrared Spectroscopy, 81, 96, 97, 110, 215, 228, 229
Free radicals, 27

G
Ganoderma lucidum, 81, 82, 95, 97, 98, 215, 216, 228, 229
GC-MS, 63, 64, 65, 66, 67, 68, 69, 70, 73, 76, 77, 79, 125, 136, 138, 141, 147

H
hyperspectral imaging, 142, 179, 180, 182, 183, 197, 199

I
ICP-MS, 201, 202, 203, 204, 206, 210, 212, 257

L
Landfill Leachate, 231, 241

M
Metabolomics, 64, 68, 70, 78, 80
Multielements, 202

O
Oat, 63, 64

P
Partial least square regression, 27
partial least squares discriminant analysis, 85, 102, 180, 215, 217, 219
Partial Least Squares Discriminant Analysis, 81, 196, 215
Peak tables, 131
Penalized Linear Discriminant Analysis, 215
Physico-Chemical Parameters, 233
Physico-Chemical Variables, 231
Polymeric Materials, 15, 25
Principal Component Analysis, 7, 15, 25, 26, 72, 96, 195, 227, 242, 250
Principal Component Discriminant Analysis, 81, 215

Q
QSAR, 5, 6, 7, 10, 29, 40, 43, 45, 46, 48, 56, 57, 58, 59, 60
quality control, 17, 113, 114, 144, 179, 180, 182, 188, 194, 197, 202, 212, 215, 217, 226, 243, 244, 254

R
Raman Microscope, 102
Raman Spectroscopy, 99, 109, 110
Reactive oxygen species, 27
Rye, 63, 64, 76

S
Spectral Acquisition, 161, 217
Suet oil, 201, 202, 243, 244, 251
SWIR Image Analysis, 195

T
TMSCN, 64, 65, 66, 76
Traditional Chinese Medicine, 81, 96, 97, 227

U
Ultrawave digestion, 202, 203, 204, 212

W
Wheat, 63, 64